CISM Courses and Lectures

The series presents lecture notes, monographs, edited works and proceedings in the field of Mechanics, Engineering, Computer Science and Applied Mathematics.
Purpose of the series is to make known in the international scientific and technical community results obtained in some of the activities organized by CISM, the International Centre for Mechanical Sciences.

International Centre for Mechanical Sciences

Courses and Lectures Vol. 553

For further volumes:
www.springer.com/series/76

Adrian Muntean · Federico Toschi
Editors

Collective Dynamics from Bacteria to Crowds

An Excursion Through Modeling, Analysis and Simulation

Springer

Editors

Adrian Muntean
Eindhoven University of Technology, The Netherlands

Federico Toschi
Eindhoven University of Technology, The Netherlands

ISSN 0254-1971
ISBN 978-3-7091-1784-2 ISBN 978-3-7091-1785-9 (eBook)
DOI 10.1007/978-3-7091-1785-9
Springer Wien Heidelberg New York Dordrecht London

All contributions have been typeset by the authors
Printed in Italy

Printed on acid-free paper

Springer is part of Springer Science+Business Media (www.springer.com)

PREFACE

The motion of individual active agents (like e.g. bacteria or ant colonies, birds flocks, etc.) give rise to fascinating large scale collective behaviours. How does this large-scale emerge from the small-scale dynamics? How does the large-scale conditions influence the dynamics of the individuals? There are many fundamental scientific questions with important practical implications that have attracted the attention of various scientific communities, ranging from logistics, theoretical biology, ecology, statistical physics and mathematics. On the one hand, one would like to understand the formation of large-scale patterns in large colonies, this may be relevant to fisheries and fishing strategies optimization. On the other hand, in crowded pedestrian flows, the behaviour of individuals display significant differences from that of undisturbed free walking. Furthermore, when panic situations occur, small microscopic (i.e. individual-level) interactions can amplify leading to macroscopic patterns (e.g. shock-like waves) that can cause jamming during evacuation but also losses of human lives.

Multiscale models in social applications combine mean-field and kinetic equations with either microscopic or macroscopic level descriptions. These are approaches of strongly increasing importance with high potential for quantitative research. Typically, individual-based models need to be accurately coarse-grained to translate the relevant microstructure information to a mesoscopic (Boltzmann-level) or to a macroscopic (continuum) level. Relevant questions include: What is the natural scaling for the averaging? How much micro-level information needs to be retained in order to capture the specific individual-level interaction responsible for the formation and propagation of the macroscopically-observed patterns (for instance, lane formation in pedestrian counterflow). What are the main microscopic interactions responsible for the macroscopic transport mechanism displacing pedestrian flows?

Within this book an attempt is made to cover a limited number of these questions with an eye on multidisciplinary approach to the topics:

J.-A. Carrillo, Y.-P. Choi and M. Hauray focus on the derivation of mean-field models for swarming proving, by means of con-

verging Wasserstein distances of empirical measures, the discrete-to-continuum passage from a first-order system of interacting particles to a continuity-like equation with nonlocal kernel. Their technique is applicable to a large class of first-order interaction models.

B. Maury treats hard congestions in models for crowd motion. The hard-core part of the interactions naturally leads to non-smooth evolution systems. The handling of the contacts translates here into suitable (quasi-)variational inequalities. Rigorous numerics show that such contacts can be quantitatively evaluated.

Pedestrians moving in the dark are modeled by A. Muntean, E. Cirillo, O. Krehel, and M. Böhm in terms of Becker-Döring interaction rules for two possible kinds of scenarios: (i) a continuum PDE model in term of measures and (ii) a lattice automaton. They show that adhering to large groups is not necessarily the right thing to do if one wishes to find invisible exits.

S. Pigolotti, R. Benzi, M. Jensen, P. Perlekar, F. Toschi discuss a model for stochastic competitions of biological species in space focusing on how the macroscopic equations for individual species density can be derived within the formalism of master equations.

F. Tesser and Ch. Doering review the non-equilibrium statistical mechanics models of reaction and interaction kinetics. Among others, they show that traditional mean-field or "mass-action" reaction kinetics theories are useful but that there are also limits to their validity.

A. Tosin reviews multiscale crowd dynamics scenarios posed in terms of conservation laws for (discrete and absolutely continuous) mass measures from a threefold perspective: modeling, solvability, and approximation.

The multiscale nature of interacting particle systems gives rise to many interesting and challenging mathematical problems. In this book, the reader will find not only a wide spectrum of multiscale analysis results (like convergence proofs), but also practically important information such as derivations of mean-field equations, methods to handle hard contacts numerically, to model group behavior, to quantitative estimate microscopic/macroscopic segregation of competing species, to quantitative understand the limits of validity of mass-action kinetics for simple reactions.

Adrian Muntean and Federico Toschi

CONTENTS

The derivation of swarming models: Mean-field limit and Wasserstein distances

José Antonio Carrillo[†], Young-Pil Choi[†], and Maxime Hauray[‡]
[†] Department of Mathematics, Imperial College London, SW7 2AZ London, United Kingdom
[‡] Centre de Mathématiques et Informatique, Université d'Aix-Marseille, Technopôle Château-Gombert, Marseille, France

Abstract These notes are devoted to a summary on the mean-field limit of large ensembles of interacting particles with applications in swarming models. We first make a summary of the kinetic models derived as continuum versions of second order models for swarming. We focus on the question of passing from the discrete to the continuum model in the Dobrushin framework. We show how to use related techniques from fluid mechanics equations applied to first order models for swarming, also called the aggregation equation. We give qualitative bounds on the approximation of initial data by particles to obtain the mean-field limit for radial singular (at the origin) potentials up to the Newtonian singularity. We also show the propagation of chaos for more restricted set of singular potentials.

1 Introduction

In the last years, we have seen the development of a great deal of different models in the biology, applied mathematics, and physics literature to describe the collective behavior of individuals. Here, individuals may mean animals (insects, fish, birds,...), bacteria, and even robots. Most of these models involve the nonlocal character of the interaction as a basic modelling pillar, see for instance Camazine, Deneubourg, Franks, Sneyd, Theraulaz, and Bonabeau (2003); Couzin, Krause, Franks, and Levin (2005); Li, Lukeman, and Edelstein-Keshet (2008); Vicsek, Czirok, Ben-Jacob, Cohen, and Shochet (1995). In fact, one of largest source of collective behavior models comes from control engineering. There, the aim is to produce a suitable control of the movement of small squads of robots in order to perform unmanned vehicle operations, for instance Perea, Gómez, and Elosequi (2009). These ideas even have been proposed to model crowd motion, including more "intelligent" particles deciding their movement based on optimization of certain quantities: time to exit from a room or a stadium, for instance Burger, Markowich, and Pietschmann (2011).

Either in social or in biological sciences, these models encounter many interesting features such as the spontaneous formation of different pattern behaviors. When we talk about patterns, we do not mean static patterns like in the study of crystals but rather dynamic patterns leading to the collective motion of the individual ensemble. For instance, two of the main collective motion patterns studied in different models are the flock and the milling behavior, see D'Orsogna, Chuang, Bertozzi, and Chayes (2006); Carrillo, D'Orsogna, and Panferov (2009); Cañizo, Carrillo, and Rosado (2010); Carrillo, Klar, Martin, and Tiwari (2010); Carrillo, Panferov, and Martin (2013). In the flock pattern, individuals achieve a consensus on the direction or orientation towards some objective, producing as a consequence a particular spatial shape showing their preferred comfort structure. This kind of swiftly moving flocks have been reported in many species although the most spectacular or bucolic ones are the bird flocks, starlings for instance. In the mill pattern, individuals arrange into a kind of vortex like motion around some point. This particular moving pattern has been observed in fish schools. Hundreds of movies can be easily accessed through internet search showing them.

There are many reasons one can argue, why such a large number of individuals react to external stimuli producing these macroscopic patterns without seemingly the presence of a leader in the swarm. Hydrodynamic enhancement, predators avoidance, social interactions, spawning survival rate, and many others have been proposed to explain this behavior in different species, see Parrish, and Edelstein-Keshet (1999).

One of the main question in describing this behavior by mathematical models is how to include the interaction between individuals. In any case, there is a consensus that the modelling starts from particle-like models as in statistical physics. These particle models are also called Individual-Based Models (IBMs) in the community. They are usually formed by a set of differential equations of Newton type (called 2nd order models) or by kinematic equations where the inertia terms are neglected (called first order models). Essentially, by assuming that the inertia term is negligible, we assume that individuals can adjust to the velocity field instantaneously, an approximation valid when their speed is not too large. In any case, these first order models were proposed in the literature derived in a phenomenological manner; see Mogilner, Edelstein-Keshet, Bent, and Spiros (2003); Mogilner and Edelstein-Keshet (1999); Parrish, and Edelstein-Keshet (1999); Topaz and Bertozzi (2004); Topaz, Bertozzi, and Lewis (2006); Eftimie, de Vries, and

Lewis (2007). The literature on first and second order models for swarming has increased exponentially fast in the last few years. Many of these models find also their origin in social sciences, where consensus or opinion formation was also described in similar grounds. Another typical ingredient in these models is some kind of noise leading to systems of SDEs. In this work, we will not discuss how to incorporate noise in these models, we refer to Bolley, Cañizo, and Carrillo (2011) and the references therein.

Most of these models are based on discrete approaches incorporating certain effects that we like to call the "first principles" of swarming. These first principles are based on modelling the "sociological behavior" of animals with very simple rules such as the social tendency to produce grouping (attraction/aggregation), the inherent minimal space they need to move without problems and feel comfortably inside the group (repulsion/collisional avoidance) and the mimetic adaptation or synchronization to a group (orientation/alignment). Even if these minimal models contain very basic rules, the patterns observed in their simulation and their complex asymptotic behavior are already very challenging from the mathematical viewpoint. The 3-zone models including attraction, repulsion, and alignment effects are classical in fish modelling; see Aoki (1982); Huth and Wissel (1992) for instance. Based on them, one can incorporate may other effects to render more realistic the outputs of the simulations and the models, see Barbaro, Taylor, Trethewey, Youseff, and Birnir (2009) for fish schools or Hemelrijk and Hildenbrandt (2008) for birds flocks. We also refer to the reader to the recent review Carrillo, Fornasier, Toscani, and Vecil (2010) about the kinetic modelling of swarming.

To the eyes of a kinetic theorist or a statistical physicist, studying such systems of ODEs when the number of individuals becomes large is doomed to fail. Dynamical system approaches are quite useful but they typically have huge problems to describe large systems of particles. A classical approach to attack the problem is to pass to a continuous description of the system. This means to go from particle descriptions to kinetic descriptions where the unknown is the particle density distribution in position-velocity (phase) space for 2nd order models or in position space for 1st order models.

Going from particle to continuum descriptions is one of the most classical problems in kinetic theory. It is at the basis of the derivation of the mother and father kinetic equations, namely: the Vlasov and the Boltzmann equations. A rigorous derivation of the Boltzmann equation from the Newtonian dynamics has only been given for short times (of the order of the average time of first collision), see Lanford (1974) Gallagher, St-Raymond, and Texier (2012). In that case, interactions between the particles are modelled by short-range potentials leading to collision kernels. The question of the derivation of the Boltzmann equation from particles with jump processes was also raised and solved by Kac (1956), and further results are given in the recent important work by Mischler and Mouhot (2013). The derivation of the Vlasov equation is well understood only for regular or not too singular potentials; see Braun and Hepp (1977); Neunzert (1984); Dobrushin (1979); Hauray and Jabin (2012). In fact, a full derivation of the Vlasov-Poisson system in 3D is also lacking. The problem of passing to the limit from particle to continuum models like the Vlasov equation is called the mean-field limit. This name just comes from the fact that the resulting equation is a kind of averaged version of the interaction between the large number of individuals. Moreover, the resulting equation gives the typical behavior of one isolated individual among all the others since they are assumed to be completely indistinguishable.

Finally, there are other famous mean-field limit equations, such as the Euler and the Navier-Stokes equations for incompressible fluids, see Marchioro and Pulvirenti (1994); Majda and Bertozzi (2002). It has been extensively used for numerical purposes that both equations in the 2D incompressible case can be derived from particle approximations, called vortex point approximations. The convergence in the viscous case has been rigorously proved for very general initial data; see Osada (1985); Founier, Hauray, and Mischler (2012). In the non-viscous case Schochet (1996) proves that particle approximations converge towards solutions of the Euler equation, but they may not converge to the good solution because of the lack of uniqueness in the Euler equation, see De Lellis and L. Székelyhidi (2009). However, in the case where the initial particles are equally spaced on a grid to approximate a smooth solution of the Euler equation, the convergence was shown in Goodman, Hou, and Lowengrub (1990). These vortex methods have been proven to be convergent and estimates of the error committed have been obtained in recent works using optimal transport techniques (Hauray (2009)) but not for the real Euler equation in 2D.

The aim of this work is to show in detail a particular example of the mean field limit in the case of first order models not covered in the previous literature. Nevertheless, we will first discuss some of these issues for 2nd order models summarizing results in Cañizo, Carrillo, and Rosado (2011); Bolley, Cañizo, and Carrillo (2011). We will also discuss that the spatial shape of the main patterns, flock and mills, are given by stationary solutions of the 1st order models. This gives another reason from a more conceptual mathematical viewpoint of reducing to 1st order models. Section 3 will be devoted to obtain the mean field limit to the so-called aggregation equation for singular potentials recovering some of the models studied in Bertozzi, Carrillo, and Laurent (2009); Bertozzi, Laurent, and Rosado (2010). Here, the idea is to assume that we have solutions of the model in better functional spaces due to the singularity of the potential, but we have to pay in terms of conditions on the initial distribution of particles (how they are distributed) in such a way that the particle solution converges to the continuum solution of the aggregation equation as $N \to \infty$. We will make use of similar arguments to Hauray (2009) to show the mean-field limit for first order swarming models with singular potentials up to the Newtonian singularity. In Section 4, we study a local existence of a unique L^p-solution for the aggregation equation. This complements the well-posedness theory in Bertozzi, Laurent, and Rosado (2010). Finally, Section 5 is devoted to show the propagation of chaos property for the aggregation equation. This property is very important from the physical relevance of the kinetic and aggregation models, since it states that one can derive the mean-field equations under quite generic randomly generated initial location of the particles. We are only able to show it for a more restricted set of singular potentials with respect to the mean-field limit.

2 The Dobrushin approach

2.1 Some Individual Based Models

As we described in the introduction, the modelling in swarming starts by introducing some particle models, IBMs in the jargon of this community, incorporating some of the basic effects: repulsion, attraction, and alignment. Let us discuss briefly some of these models, starting with the ones that have recently attracted more attention due to their simplicity while having a rich mathematical structure and pattern formation. One of these models was introduced by the UCLA group in D'Orsogna, Chuang, Bertozzi, and Chayes (2006) and it consists of Newton-like equations where all the effect of repulsion and attraction is encoded via a pairwise potential $W : \mathbb{R}^d \to \mathbb{R}$. A popular choice for the interaction potential W is the Morse potential given by

$$W(x) = -C_A e^{-|x|/\ell_A} + C_R e^{-|x|/\ell_R}, \tag{2.1}$$

where C_A, C_R and ℓ_A, ℓ_R are the strengths and the typical lengths of attraction and repulsion, respectively. They are chosen for having biologically reasonable potentials with $C = C_R/C_A > 1$ and $\ell_R/\ell_A < 1$, see Carrillo, Panferov, and Martin (2013) for other nice choices of the interaction potentials and a deeper discussion on the issue of biologically relevant interaction potentials. Apart from this, the other effect included is the tendency of the particles to travel asymptotically at a fixed speed as in Levine, Rappel, and Cohen (2000). Consequently, a term producing a balance between self-propulsion and friction is introduced imposing an asymptotic speed to the particles (if other effects are ignored), but it does not influence the orientation vector. The resulting ODE system reads as:

$$\begin{cases} \dfrac{dx_i}{dt} = v_i, & (i = 1, \dots, N), \\ \dfrac{dv_i}{dt} = (\alpha - \beta\,|v_i|^2)v_i - \dfrac{1}{N}\displaystyle\sum_{j\neq i} \nabla W(|x_i - x_j|), & (i = 1, \dots, N). \end{cases}$$

where α, β are nonnegative parameters, determining the asymptotic speed of particles given by $\sqrt{\alpha/\beta}$. Here, the potential has been scaled depending on the mass of each particle as in Carrillo, D'Orsogna, and Panferov (2009) and in such a way that the effect of the potential per particle diminishes while the energy is of constant order as the number of particles N diverges. This scaling is the so-called mean-field scaling, see the introduction of Bodnar and Velazquez (2012) for a nice discussion of the different scalings in first order models.

Another popular IBM including only the alignment effect is the so-called Cucker and Smale (2007) model. Each individual in the swarm changes its velocity vector based on the other individuals by adjusting/averaging their relative velocity with all the others. This averaging is weighted in such a way that closer individuals have more influence than further ones. For a system with N individuals the Cucker-Smale model reads as

$$\begin{cases} \dfrac{dx_i}{dt} = v_i, \\ \dfrac{dv_i}{dt} = \dfrac{1}{N}\displaystyle\sum_{j=1}^{N} w_{ij}\,(v_j - v_i)\,, \end{cases}$$

with the *communication rate* $w(x)$ given by:

$$w_{ij} = w(x_i - x_j) = \frac{1}{(1 + |x_i - x_j|^2)^{\gamma}}\,,$$

for some $\gamma \geq 0$.

Associated to the above models, one can formally write the expected Vlasov-like kinetic equations as $N \to \infty$, see for instance Carrillo, D'Orsogna, and Panferov (2009), leading to

$$\partial_t f + v \cdot \nabla_x f - (\nabla W * \rho) \cdot \nabla_v f + \mathrm{div}_v((\alpha - \beta|v|^2)vf) = 0, \qquad (2.2)$$

where ρ represents the macroscopic *density* of f:

$$\rho(t,x) := \int_{\mathbb{R}^d} f(t,x,v)\,dv \quad \text{for } t \geq 0, \quad x \in \mathbb{R}^d.$$

The Cucker-Smale particle model leads to the following kinetic equation:

$$\frac{\partial f}{\partial t} + v \cdot \nabla_x f = \nabla_v \cdot [\xi[f]\, f]\,, \tag{2.3}$$

where $\xi[f](x,v,t) = (H * f)\,(x,v,t)$, with $H(x,v) = w(x)v$ and $*$ standing for the convolution in both position and velocity (x and v). We refer to Cucker and Smale (2007); Ha and Tadmor (2008); Ha and Liu (2009); Carrillo, Fornasier, Rosado, and Toscani (2010) for further discussion about this model and qualitative properties.

Moreover, quite general models incorporating the three effects previously discussed with additional ingredients, such as vision cones or topological interactions, have been considered in Carrillo, Fornasier, Toscani, and Vecil (2010); Li, Lukeman, and Edelstein-Keshet (2008); Agueh, Illner, and Richardson (2011); Albi and Pareschi (2013); Haskovec (2013). In particular Li, Lukeman, and Edelstein-Keshet (2008) consider that the N individuals follow the system:

$$\begin{cases} \dfrac{dx_i}{dt} = v_i, \\[2mm] \dfrac{dv_i}{dt} = F_i^A + F_i^I, \end{cases} \tag{2.4}$$

where F_i^A is the self-propulsion generated by the ith-individual, while F_i^I is due to interaction with the others. The interaction with other individuals can be generally modeled as:

$$F_i^I = F_i^{I,x} + F_i^{I,v} = \sum_{j=1}^{N} g_\pm(|x_i - x_j|)\frac{x_j - x_i}{|x_i - x_j|} + \sum_{j=1}^{N} h_\pm(|v_i - v_j|)\frac{v_j - v_i}{|v_i - v_j|}.$$

Here, g_+ and h_+ (g_- and h_-) are chosen when the influence comes from the front (behind), i.e., if $(x_j - x_i) \cdot v_i > 0$ (< 0); choosing $g_+ \neq g_-$ and $h_+ \neq h_-$ means that the forces from particles in front and those from particles behind are different. The sign of the functions $g_\pm(r)$ encodes the short-range repulsion and long-range attraction for particles in front of (+) and behind (-) the ith-particle. Similarly, $h_+ > 0$ (< 0) implies that the velocity-dependent force makes the velocity of particle i get closer to (away from) that of particle j.

Some of these models, for instance Agueh, Illner, and Richardson (2011); Albi and Pareschi (2013); Haskovec (2013), include sharp boundaries for the vision cone or for the interaction with the nearest neighbors. As we shall see later, these are typical situations in which the mean-field limit for general measures will not work. By sharp boundaries we mean that the functions involved in the kernels such as $w(x)$, $g_\pm$, or $h_\pm$ are given by characteristic functions on sets depending on the location/velocity of the agent.

2.2 First-order models: Aggregation Equation

In this work, the objective is to show how to obtain the continuum limits of these particle models in a simpler situation than the ones in the previous section. However, at the same time we will allow for more singular kernels. We will showcase these tools in the case of the so-called aggregation equation. Let us assume that we have just particles interacting through the pairwise potential $W(x)$. Assuming that the variations of the velocity and speed are much smaller than spatial variations, see Mogilner and Edelstein-Keshet (1999), then one can neglect the inertia term in Newton's equation to deduce that

$$\frac{dX_i}{dt} = -\sum_{j\neq i} \nabla W(X_i - X_j) \text{ in the } N \to \infty \text{ limit} \Rightarrow \begin{cases} \dfrac{\partial \rho}{\partial t} + \operatorname{div}(\rho u) = 0 \\ u = -\nabla W * \rho \end{cases}. \tag{2.5}$$

Another reason to study this first order equation is that the stationary states of the first order model determine the spatial shape of the flock solutions to the second order models, see Carrillo, Panferov, and Martin (2013).

Let us note that some of the difficulties to overcome are already in this model. Next subsection is devoted to review the classical Dobrushin strategy for the mean-field limit when all functions involved in the model are smooth enough. This strategy applies to the aggregation equation for $\mathcal{C}^2(\mathbb{R}^d)$ smooth potential with at most quadratic growth at infinity by following the same argument as in Theorem 2.4 below. This argument was detailed in a nice summer school notes in Golse (2003). The goal of this chapter is to show how to deal with more singular potentials. The main message is that in order to obtain the mean-field limit, whose precise statement is given later on, you need to impose certain conditions on the approximation of the initial data avoiding the possible singularities (collisions) in finite time of the particles. We will elaborate on this at the begining of next section. In order to deal with these questions, it is quite convenient to work with transport distances between probability measures that we quickly review next.

2.3 Basic tools in transport distances

In this subsection, we present several definitions of Wasserstein distances and their properties.

Definition 2.1. (Wasserstein p-distance) Let ρ_1, ρ_2 be two Borel probability measures on $\mathbb{R}^d$. Then the Euclidean Wasserstein distance of order $1 \le p < \infty$ between ρ_1 and ρ_2 is defined as

$$d_p(\rho_1, \rho_2) := \inf_\gamma \left(\int_{\mathbb{R}^d \times \mathbb{R}^d} |x-y|^p \, d\gamma(x,y) \right)^{1/p},$$

and, for $p = \infty$ (this is the limiting case, as $p \to \infty$),

$$d_\infty(\rho_1, \rho_2) := \inf_\gamma \left(\sup_{(x,y) \in \text{supp}(\gamma)} |x-y| \right),$$

where the infimum runs over all transference plans, i.e., all probability measures γ on $\mathbb{R}^d \times \mathbb{R}^d$ with marginals ρ_1 and ρ_2 respectively,

$$\int_{\mathbb{R}^d \times \mathbb{R}^d} \phi(x) d\gamma(x,y) = \int_{\mathbb{R}^d} \phi(x) \rho_1(x) dx,$$

and

$$\int_{\mathbb{R}^d \times \mathbb{R}^d} \phi(y) d\gamma(x,y) = \int_{\mathbb{R}^d} \phi(y) \rho_2(y) dy,$$

for all $\phi \in \mathcal{C}_b(\mathbb{R}^d)$.

We also remind the definition of the push-forward of a measure by a mapping in order to give the relation between Wasserstein distances and optimal transportation.

Definition 2.2. Let ρ_1 be a Borel measure on $\mathbb{R}^d$ and $\mathcal{T} : \mathbb{R}^d \to \mathbb{R}^d$ be a measurable mapping. Then the push-forward of ρ_1 by $\mathcal{T}$ is the measure ρ_2 defined by

$$\rho_2(B) = \rho_1(\mathcal{T}^{-1}(B)) \quad \text{for} \quad B \subset \mathbb{R}^d,$$

and denoted as $\rho_2 = \mathcal{T} \# \rho_1$.

The set of probability measures with bounded moments of order p, denoted by $\mathcal{P}_p(\mathbb{R}^d)$, $1 \le p < \infty$, is a complete metric space endowed with the p-Wassertein distance d_p, see Villani (2003). We refer to Givens and Shortt (1984); McCann (2006) for more details in the case of the d_∞ distance.

Remark 2.3. The definition of $\rho_2 = \mathcal{T} \# \rho_1$ is equivalent to

$$\int_{\mathbb{R}^d} \phi(x) \, d\rho_2(x) = \int_{\mathbb{R}^d} \phi(\mathcal{T}(x)) \, d\rho_1(x),$$

for all $\phi \in \mathcal{C}_b(\mathbb{R}^d)$. Given a probability measure with bounded p-th moment ρ_0, consider two measurable mappings $X_1, X_2 : \mathbb{R}^d \to \mathbb{R}^d$, then the following inequality holds.

$$d_p^p(X_1\#\rho_0, X_2\#\rho_0) \leq \int_{\mathbb{R}^d\times\mathbb{R}^d} |x-y|^p d\gamma(x,y) = \int_{\mathbb{R}^d} |X_1(x)-X_2(x)|^p d\rho_0(x).$$

Here, we used as transference plan $\gamma = (X_1 \times X_2)\#\rho_0$ in Definition 2.1.

2.4 A quick review of the classical Dobrushin result

Under smoothness assumptions on the ingredient functions of the swarming models, one can use adaptations of the classical result of Dobrushin (1979) to obtain what is called the mean-field limit equation for general particle approximations of any initial measure. These arguments are classical in kinetic theory and were also introduced in Braun and Hepp (1977); Neunzert (1984), making use of the bounded Lipschitz distance, and reviewed in Spohn (1991); Villani (2002), see also Sznitman (1991); Méléard (1996) for the case with noise. The bounded Lipschitz distance or dual $W^{1,\infty}$-norm is equivalent to the Wasserstein distance d_1 for compactly supported measures. This strategy works as soon as the velocity field defining the characteristics of the model is a bounded and globally Lipschitz function whose dependence on the measure itself is Lipschitz continuous in the d_1 sense. These ideas were improved to allow for locally Lipschitz velocity fields for compactly supported initial measures in Cañizo, Carrillo, and Rosado (2011) and for suitable decay conditions at infinity and with noise in Bolley, Cañizo, and Carrillo (2011). With these techniques one can include quite general kinetic models for swarming in this well-posedness theory.

Let us introduce some notation for this section: $\mathcal{A} = \mathcal{P}_c(\mathbb{R}^d \times \mathbb{R}^d)$ denotes the subset of $\mathcal{P}(\mathbb{R}^d \times \mathbb{R}^d)$ consisting of measures of compact support in $\mathbb{R}^d \times \mathbb{R}^d$. On the other hand, we consider the set of functions $\mathcal{B} := \mathrm{Lip}_{loc}(\mathbb{R}^d \times \mathbb{R}^d, \mathbb{R}^d)$, which in particular are locally Lipschitz with respect to (x,v). B_R will denote the ball centered at 0 of radius R in $\mathbb{R} \times \mathbb{R}$.

Let us consider general operators from measures to vector fields, $\mathcal{H}[\cdot] : \mathcal{A} \to \mathcal{B}$, satisfying the following hypotheses: for any $R_0 > 0$ and $f, g \in \mathcal{A}$ such that $\operatorname{supp} f \cup \operatorname{supp} g \subseteq B_{R_0}$, there exists some ball $B_R \subset \mathbb{R}^d \times \mathbb{R}^d$ and a constant $C = C(R, R_0) > 0$, such that

$$\|\mathcal{H}[f] - \mathcal{H}[g]\|_{L^\infty(B_R)} \leq C\, d_1(f,g), \tag{2.6}$$

$$\mathrm{Lip}_R(\mathcal{H}[f]) \leq C, \qquad \|\mathcal{H}[f]\|_{L^\infty(B_R)} \leq C. \tag{2.7}$$

Here, $\mathrm{Lip}_R(\cdot)$ denotes the Lipschitz constant of a function in B_R.

Given $f \in \mathcal{C}([0,T],\mathcal{P}_c(B_{R_0}))$, and for any initial condition $(X^0,V^0) \in \mathbb{R}^d \times \mathbb{R}^d$, the following system of ordinary differential equations has a unique locally defined solution

$$\frac{d}{dt}X = V, \qquad X(0) = X^0 \tag{2.8a}$$

$$\frac{d}{dt}V = \mathcal{H}[f(t)](X,V), \qquad V(0) = V^0. \tag{2.8b}$$

We will additionally require that the solutions to that system are "global". More precisely, we assume that for any $R_0, T > 0$, there exists $R > 0$ such that $(X(t),V(t)) \in B_R$ for all $t \in [0,T]$ and all $(X^0,V^0) \in B_{R_0}$. Of course, this is a requirement that has to be checked for every particular model. We prefer to give a general condition which reduces the problem of existence and stability to the simpler one of existence of the ODEs. Under the above conditions, the existence and uniqueness of associated transport equation

$$\partial_t f + v \cdot \nabla_x f - \nabla_v \cdot [\mathcal{H}[f]f] = 0. \tag{2.9}$$

was obtained in Cañizo, Carrillo, and Rosado (2011) to which we refer for full details. In Cañizo, Carrillo, and Rosado (2011), the interactions $\mathcal{H}[f] = (\alpha - \beta|v|^2)v - \nabla W * \rho$ and $\mathcal{H}[f] = H * f$ corresponding to (2.2) and (2.3), respectively, and

$$\mathcal{H}[f] = F_A(x,v) + G(x) * \rho + H(x,v) * f,$$

with F_A, G and H given functions satisfying suitable hypotheses, such that the kinetic equation (2.9) corresponds to the model (2.4) are investigated.

Theorem 2.4. *Given an operator $\mathcal{H}[\cdot] : \mathcal{A} \to \mathcal{B}$ satisfying Hypotheses (2.6) and (2.7) for which the characteristics (2.8a)-(2.8b) are globally well-defined, and f_0 a measure on $\mathbb{R}^d \times \mathbb{R}^d$ with compact support. There exists a solution f on $[0,+\infty)$ to equation (2.9) with initial condition f_0. In addition,*

$$f \in \mathcal{C}([0,+\infty);\mathcal{P}_c(\mathbb{R}^d \times \mathbb{R}^d)) \tag{2.10}$$

and there is some increasing function $R = R(T)$ such that for all $T > 0$,

$$\operatorname{supp} f_t \subseteq B_{R(T)} \subseteq \mathbb{R}^d \times \mathbb{R}^d \quad \text{for all } t \in [0,T]. \tag{2.11}$$

This solution is unique among the family of solutions satisfying (2.10) and (2.11). Moreover, given any other initial data $g_0 \in \mathcal{P}_c(\mathbb{R}^d \times \mathbb{R}^d)$ and g its corresponding solution, there exists a strictly increasing function $r(t)$:

$[0,\infty) \to \mathbb{R}_0^+$ *with* $r(0) = 1$ *depending only on* $\mathcal{H}$ *and the size of the support of* f_0 *and* g_0*, such that*

$$d_1(f_t, g_t) \le r(t)\, d_1(f_0, g_0), \quad t \ge 0.$$

The stability theorem 2.4 gives in particular a rigorous derivation of the kinetic equation (2.9) from the large particle limit of the system of ordinary differential equations. This is the exact statement of the mean-field limit for general measures as initial data. Let us consider the system of ordinary differential equations:

$$\dot{x}_i = v_i, \qquad i = 1, \dots, N, \tag{2.12a}$$

$$\dot{v}_i = \sum_{j \neq i} m_j \mathcal{H}[f^N(t)](x_i, v_i), \qquad i = 1, \dots, N. \tag{2.12b}$$

where $m_1, \dots, m_N \ge 0$ and $\sum_i m_i = 1$ and f^N is defined next. Under the conditions of Theorem 2.4, we first notice that if $x_i, v_i : [0,T] \to \mathbb{R}^d$, for $i = 1, \dots, N$, are a solution to the system (2.12), then the function $f^N : [0,T] \to \mathcal{P}_c(\mathbb{R}^d \times \mathbb{R}^d)$ given by

$$f_t^N := \sum_{i=1}^N m_i\, \delta_{(x_i(t), v_i(t))} \tag{2.13}$$

is the solution to (2.9) with initial condition

$$f_0^N = \sum_{i=1}^N m_i\, \delta_{(x_i(0), v_i(0))}. \tag{2.14}$$

In fact, the solution (2.13) is called the empirical measure associated to the system of ODEs (2.12). We finally write the full statement of the mean-field limit in the Dobrushin strategy.

Corollary 2.5. *Given* $f_0 \in \mathcal{P}_c(\mathbb{R}^d \times \mathbb{R}^d)$ *and* $\mathcal{H}[f]$ *satisfying the conditions of Theorem 2.4, take a sequence of* f_0^N *of measures of the form* (2.14) *(with* m_i*,* $x_i(0)$ *and* $v_i(0)$ *possibly varying with* N*), in such a way that*

$$\lim_{N \to \infty} d_1(f_0^N, f_0) = 0.$$

Consider f_t^N *the empirical measure associated to the solution of the system* (2.12) *with initial conditions* $x_i(0)$*,* $v_i(0)$*. Then,*

$$\lim_{N \to \infty} d_1(f_t^N, f_t) = 0, \tag{2.15}$$

for all $t \geq 0$*, where* $f = f(t, x, v)$ *is the unique measure solution to eq.* (2.9) *with initial data* f_0.

This section can be directly applied to the models recently introduced in Agueh, Illner, and Richardson (2011) to account for vision cones and braking/acceleration of individuals and those in Albi and Pareschi (2013); Haskovec (2013) to include topological (nearest neighbours) interactions once the parameter functions are smoothed out to avoid sharp boundaries.

Summarizing this subsection, under suitable smoothness of the parameters involved in the swarming models, the empirical measures are solutions themselves of the Vlasov-like kinetic equation (2.9). Thus, a stability result in d_1 with respect to the initial data is enough to conclude the mean-field limit. Let us consider one of the particular examples in subsection 2.1, the model introduced in D'Orsogna, Chuang, Bertozzi, and Chayes (2006) with the Morse potential (2.1). This potential does not satisfy the smoothness assumption in Theorem 2.4. In principle, one cannot expect to have a mean-field result for general measures as initial data and for general approximations by particles. In fact, we do not have a well-posedness theory for such initial data in those cases. However, one can develop well-posedness theories in better functional spaces, say $L^1 \cap L^p(\mathbb{R}^d \times \mathbb{R}^d)$ for the initial data and then impose suitable conditions to the distribution of the approximated particles initially to be able to conclude the mean-field limit (2.15). This is the strategy that have been followed in Hauray and Jabin (2012) for the classical Vlasov equation and in Hauray (2009) for Euler-like equations in fluid mechanics. The rest of this work is to show this technique applied to the first order model introduced in Subsection 2.2.

3 Mean-Field Limit for the Aggregation Equation

Now, we analyse the mean-field limit of the first order model (2.5) for swarming introduced in the previous section. More precisely, we will study sufficient conditions on the initial distribution of particles for the convergence of a particle system towards the aggregation equation. This mean-field limit model consists of the continuity equation for the probability density of individuals $\rho(x,t)$ at position $x \in \mathbb{R}^d$ and time $t > 0$ given by:

$$\begin{cases} \partial_t \rho + \nabla \cdot (\rho u) = 0, & t > 0, \quad x \in \mathbb{R}^d, \\ u(t,x) := -\nabla W * \rho, & t > 0, \quad x \in \mathbb{R}^d, \\ \rho(0,x) := \rho^0(x), & x \in \mathbb{R}^d, \end{cases} \tag{3.16}$$

where $u(x,t)$ is a velocity field non-locally computed in terms of the density of individuals.

As an approximation by particles of the aggregation equation (3.16), we consider the following ODE system:

$$\begin{cases} \dot{X}_i(t) = -\sum_{j\neq i} m_j \nabla W(X_i(t) - X_j(t)), \\ X_i(0) = X_i^0, \qquad i = 1, \ldots, N. \end{cases} \tag{3.17}$$

Here, $\{X_i\}_{i=1}^N$ and $\{m_i\}_{i=1}^N$ are the positions and weights of i-th particles, respectively. We define the associated empirical distribution $\mu_N(t)$ as

$$\mu_N(t) = \sum_{i=1}^N m_i \delta_{X_i(t)}, \quad \sum_{i=1}^N m_i = \int_{\mathbb{R}^d} \rho_0(x)dx = 1, \tag{3.18}$$

with $m_i > 0$, $i = 1, \ldots, N$. As long as two particles (or more) do not collide, and if we set $\nabla W(0) = 0$ (arbitrarily if there is a singularity), then μ_N satisfies (3.16) in the sense of distributions, i.e., $\mu_N(t)$ and $\rho(t)$ satisfy the same equation. In this framework, the convergence:

"$\mu_N^0 \rightharpoonup \rho^0$ *weakly-$*$ as measures* $\Longrightarrow \mu_N(t) \rightharpoonup \rho(t)$ *weakly-$*$ as measures for small time or for every time?*"

is a natural question. If the answer is yes, we say that the continuity equation (3.16) is the mean-field limit of the particle approximation (3.17). In other words, we can say that the continuum nonlocal equation (3.16) has been rigorously derived from particle systems.

Because of the singularity in the interaction force, the natural transport distance to use is the one induced by the d_∞-topology. In fact, the d_∞ allows to take advantage of the fact that the singularity in the interaction force is localized. This seems more difficult to use with other distance like d_p, $p < \infty$. Remark that this distance also allows to understand linearized stability of particle systems around singular steady state measures with a ring shape in first order aggregation models, see Balagué, Carrillo, Laurent, and Raoul (2013b); Kolokonikov, Sun, Uminsky, and Bertozzi (2011). Actually, a local perturbation of the dynamical system (3.17) keeping the number of particles fixed is obtained by transporting the particle to other locations nearby. One could even allow for splitting of the mass into different particles, but all of them located in a local neighborhood of the unperturbed particle positions. Certainly, sending a small portion of mass very far away from the location of one particle is not a d_∞-perturbation of the atomic measure but it is a d_p small perturbation for all $1 \leq p < \infty$. These ideas have also recently been used in Balagué, Carrillo, Laurent, and Raoul (2013b) to study local minimizers of the energy functional associated to (3.16).

Another issue to cope with is that we are dealing with particle systems whose characteristics may lead to collisions in finite time. Therefore, we will be able to obtain meaningful results only on intervals in which collisions are avoided (although in some particular cases we can allow collisions).

We next introduce several notations that are used throughout the rest of this work to compare the distance between a solution $\rho(t)$ of the continuum aggregation equation (3.16) and the empirical measure $\mu_N(t)$ defined by (3.18) associated to a solution $\{X_i\}_{i=1}^N$ of the particle system (3.17). The main two quantities appearing in this comparison are the d_∞-distance between $\rho(t)$ and $\mu_N(t)$, and the minimum inter-particle distance:

$$\eta(t) := d_\infty(\mu_N(t), \rho(t)), \quad \eta_m(t) := \min_{1\le i\neq j\le N} \left(|X_i(t) - X_j(t)|\right), \tag{3.19}$$

with $\eta^0 := \eta(0)$ and $\eta_m^0 := \eta_m(0)$. Our strategy does not take advantage, as we do not know how, of the repulsive or attractive character of the potentials, the proof being equal for both cases.

A theory of well-posedness for measure solutions has been obtained for the aggregation equation (3.16) allowing collision of particles in finite time in Carrillo, Di Francesco, Figalli, Laurent, and Slepčev (2011); Carrillo, Di Francesco, Figalli, Laurent, and Slepčev (2012). In these works, the potential is assumed to be smooth except at the origin, where the allowed singularity cannot be worse than Lipschitz and the potential has to be λ-convex, see Carrillo, Di Francesco, Figalli, Laurent, and Slepčev (2011) for details. This convexity allows for attractive at the origin potentials, but not repulsive, with local behaviors of the form $|x|^b$ with $1 \le b < 2$. In these works, the essential tools that allow to get the mean-field limit for more singular potentials that quadratic are based on gradient flows in the Wasserstein distance d_2 sense as in Ambrosio, Gigli, and Savaré (2005). The additional dissipation in the system of the natural Liapunov functional given by the total interaction energy is crucial to get the mean field limit for general measures for a potential behaving locally at 0 like $W(x) \simeq |x|$, for instance for the attractive Morse potential $W(x) = 1 - e^{-|x|}$.

In this work, we want to allow for more singular potentials at the origin as in Bertozzi, Carrillo, and Laurent (2009); Bertozzi, Laurent, and Rosado (2010), and thus we need to work with solutions in better functional spaces. More precisely, we will work with solutions of the aggregation equation (3.16) in $L^\infty(0,T;(L^1\cap L^p)(\mathbb{R}^d))$ with $1 \le p \le \infty$ to be determined depending on the singularity of the potential. We will use the notation

$$\|\rho\|_{(L^1\cap L^p)(\mathbb{R}^d)} := \|\rho\|_1 + \|\rho\|_p, \quad \|\rho\| := \|\rho\|_{L^\infty(0,T;(L^1\cap L^p)(\mathbb{R}^d))},$$

where $\|\rho\|_p$ denotes the $L^p(\mathbb{R}^d)$ norm of ρ, $1 \le p \le \infty$.

In order to make sense of solutions to (3.16), we need the following assumptions on the interaction potential: we first fix $W(0) = 0$ by definition, even if W is singular at the origin, and

$$|\nabla W(x)| \leq \frac{C}{|x|^{\alpha}}, \quad \text{and} \quad |D^2 W(x)| \leq \frac{C}{|x|^{1+\alpha}}, \quad \forall\, x \in \mathbb{R}^d \backslash \{0\}, \tag{3.20}$$

for $-1 \leq \alpha < d-1$. Note that due to the assumptions on W, we can always find $1 < p < \infty$ such that $(\alpha+1)p' < d$, and thus ∇W belongs to $\mathcal{W}^{1,p'}_{loc}(\mathbb{R}^d)$.

Our results also apply with minor modifications for interaction potentials of the form $W := W_1 + W_2$, with W_1 satisfying assumptions (3.20), and ∇W_2 being a global Lipschitz function, or even more general satisfying a one-sided Lipschitz (or convexity) condition $y \cdot D^2 W_2(x) y \leq C|y|^2$ for all $y \in \mathbb{R}^d$. This last generalization is important because it is satisfied if $W_2 = c|x|^a, (0 \leq a \leq 2)$ with c positive. So that any repulsive-attractive potential W, see Balagué, Carrillo, Laurent, and Raoul (2013a,b) for a definition, such that $W(x) \simeq -|x|^b/b$ locally at x near the origin, satisfies assumptions (3.20) locally with $\alpha = 1 - b$. Therefore, our mean-field limit results apply to locally repulsive potentials with exponent range $2 - d < b < a \leq 2$ and without much restriction on the attractive part at $+\infty$, i.e., $a > 0$. We will discuss further on localizing assumptions (3.20) at the end of this section. Finally, we cannot apply our techniques to the Newtonian singularity (Bertozzi, Laurent, and Léger (2012)) being the limiting case of our strategy as it was the case for the Euler-like models in fluid mechanics studied in Hauray (2009).

We next summarize the results on the existence and uniqueness of solutions to the aggregation equation (3.16). For the local well-posedness of solutions to equation (3.16), we refer to Bertozzi and Laurent (2007); Bertozzi, Carrillo, and Laurent (2009); Bertozzi, Laurent, and Rosado (2010); Laurent (2007). In particular, unique solutions for the system (3.16) were obtained in Bertozzi, Laurent, and Rosado (2010) with second moment bounded initial data. More precisely, Bertozzi et al. (Bertozzi, Laurent, and Rosado, 2010, Theorem 1.1) showed that if $\nabla W \in \mathcal{W}^{1,p'}(\mathbb{R}^d)$ and $\rho^0 \in L^p(\mathbb{R}^d) \cap \mathcal{P}_2(\mathbb{R}^d)$, then there exists $T^* > 0$ and a unique nonnegative solution to (3.16) satisfying

$$\rho \in \mathcal{C}([0,T^*], (L^1 \cap L^p)(\mathbb{R}^d)) \cap \mathcal{C}^1([0,T^*], \mathcal{W}^{-1,p}(\mathbb{R}^d)).$$

Unfortunately, one can not directly apply those results for potentials satisfying assumptions (3.20). We will compliment the results in Bertozzi, Laurent, and Rosado (2010) to show the local existence of a unique solution to the system (3.16) with the interaction potential function W satisfying

(3.20) in Section 4. We prefer to postpone the well-posedness theory in order to emphasize the mean-field limit result contained in the following theorem, whose proof follows the strategy in Hauray (2009).

Theorem 3.1. *Suppose the kernel W satisfies* (3.20), *and let ρ be a solution to the system* (3.16) *up to time $T > 0$, such that $\rho \in L^\infty(0,T;(L^1 \cap L^p)(\mathbb{R}^d)) \cap \mathcal{C}([0,T],\mathcal{P}_1(\mathbb{R}^d))$, with initial data $\rho^0 \in (\mathcal{P}_1 \cap L^p)(\mathbb{R}^d)$, $0 \le \alpha < -1 + d/p'$, and $1 < p \le \infty$. Furthermore, we assume μ_N^0 converges to ρ^0 for the distance d_∞ as the number of particles N goes to infinity, i.e.,*

$$d_\infty(\mu_N^0, \rho^0) \to 0 \quad as \quad N \to \infty,$$

and that the initial quantities η^0, η_m^0 satisfy

$$\lim_{N\to\infty} \frac{(\eta^0)^{d/p'}}{(\eta_m^0)^{1+\alpha}} = 0. \tag{3.21}$$

Then, for N large enough the particle system (3.17) *is well-defined up to time T, in the sense that there is no collision between particles before that time, and moreover*

$$\mu_N(t) \rightharpoonup \rho(t) \quad weakly\text{-}* \text{ as measures as} \quad N \to \infty, \quad \text{for all} \quad t \in [0,T].$$

Remark 3.2. Let us first discuss the assumptions on the initial data in Theorem 3.1. The mean-field limit is valid for particular approximations μ_N^0 of ρ^0, that is, for well chosen particle approximations of the initial data. In fact, a procedure to construct initial atomic measures approximating the initial condition in the sense of (3.21) is the following: define a regular mesh of size ε and approximate ρ^0 by a sum of Dirac masses μ_N^0 located at the center of the cells such that the mass at each particle is exactly equals to the mass of ρ^0 contained in the associated cell. In that case, we have $\eta^0 \sim \varepsilon$ and $\eta_m^0 \sim \varepsilon$ (for the last condition we need that the mesh has some regularity). In that case, the assumption (3.21) is automatically fulfilled since $(1+\alpha)p' < d$. Notice that no bound on the masses m_i of the particles is required.

Proof of Theorem 3.1. The proof of Theorem 3.1 is divided into three steps:

- In Step A, we estimate the growth of the d_∞ Wasserstein distance between the continuum and the discrete solutions η that involves η itself and η_m in the form:

$$\frac{d\eta}{dt} \le C\eta\|\rho\| \left(1 + \eta^{d/p'} \eta_m^{-(1+\alpha)}\right). \tag{3.22}$$

- In Step B, we estimate the decay of the minimum inter-particle distance η_m, which also involves the terms η and η_m in the form:

$$\frac{d\eta_m}{dt} \geq -C\eta_m \|\rho\| \left(1 + \eta^{d/p'} \eta_m^{-(1+\alpha)}\right). \tag{3.23}$$

- In Step C, under the assumption of the initial approximation (3.21), we combine (3.22) and (3.23) to conclude the desired result.

Step A.- We first introduce the flows generated by the two velocity fields: $u(x,t) = -\nabla W * \rho$ and $u_N := -\nabla W * \mu_N$. Let us remark that the convolution in the definition of u_N is just a notation for the right-hand side of (3.17) since the convolution of a Dirac Delta with a (possibly) singular potential is not well-defined. These flows $\Psi_N, \Psi : \mathbb{R}_+ \times \mathbb{R}_+ \times \mathbb{R}^d \to \mathbb{R}^d$ are defined as solutions of

$$\begin{cases} \dfrac{d}{dt}(\Psi(t;s,x)) = u(t;s,\Psi(t;s,x)), \\ \Psi(s;s,x) = x, \end{cases} \tag{3.24}$$

for all $s,t \in [0,T]$, and

$$\begin{cases} \dfrac{d}{dt}(\Psi_N(t;s,x)) = u_N(t;s,\Psi_N(t;s,x)), \\ \Psi_N(s;s,x) = x, \end{cases} \tag{3.25}$$

for all $s,t \in [0,T_0^N]$. Notice that the solution $X_i(t)$ to the system (3.17) is well-defined and continuous by the Cauchy-Lipschitz theorem as long as there is no collision between particles. Since $\eta_m^0 > 0$, there exists $T_0^N > 0$ such that $\eta_m(t) > 0$ for $t \in [0,T_0^N]$ by continuity. Then the flow map $\Psi_N(t;s,x)$ solution to (3.25) is well-defined for $t,s \in [0,T_0^N]$. Now, let us check that the flow for the solution associated to the continuum equation in (3.24) is well-defined. Assumptions (3.20) imply that

$$|\nabla W(x) - \nabla W(y)| \leq \frac{2|x-y|}{\min(|x|,|y|)^{\alpha+1}}. \tag{3.26}$$

One can see this by integrating along a straight line joining x and y but avoiding the singularity using a small circle if needed, see Hauray (2009). The estimate (3.26) implies that the velocity field is Lipschitz continuous with respect to the spatial variable. Actually, one can estimate it as

$$
\begin{aligned}
|u(t,x)-u(t,y)| &\le \int_{\mathbb{R}^d} |\nabla W(x-z)-\nabla W(y-z)|\rho(t,z)\,dz \\
&\le 2|x-y|\int_{\mathbb{R}^d} \frac{1}{\min(|x-z|,|y-z|)^{\alpha+1}}\rho(t,z)\,dz \\
&\le 4|x-y| \sup_{x\in\mathbb{R}^d}\int_{\mathbb{R}^d} \frac{1}{|x-z|^{\alpha+1}}\rho(t,z)\,dz\,.
\end{aligned}
$$

Now, splitting the last integral into the near- and far-field sets $\mathcal{A} := \{z : |x-z| \ge 1\}$ and $\mathcal{B} := \mathbb{R}^d - \mathcal{A}$ and estimating the two terms, we deduce

$$
\begin{aligned}
\int_{\mathbb{R}^d} \frac{1}{|x-z|^{\alpha+1}}\rho(t,z)\,dz &\le \|\rho(t)\|_1 + \left(\int_{\mathcal{B}} \frac{1}{|x-y|^{(1+\alpha)p'}}dy\right)^{1/p'} \|\rho(t)\|_p \\
&\le C\|\rho\|\,, \qquad (3.27)
\end{aligned}
$$

for all $x \in \mathbb{R}^d$ due to the assumption $(1+\alpha)p' < d$. Putting together previous inequalities, we get the desired Lipschitz continuity of the velocity field with respect to x, which is moreover uniform in time. A similar estimate using (3.20) shows that the velocity field is bounded, and then the flow Ψ in (3.24) is well-defined. Our first aim is to find an expansion of the velocity of the d_∞ Wasserstein distance. The idea is similar to the evolution of the euclidean Wassertein distance in Carrillo, McCann, and Villani (2003, 2006); Otto (2001). Fixed $0 \le t_0 < \min(T, T_0^N)$ and choose an optimal transport map for d_∞ denoted by $\mathcal{T}^0$ between $\rho(t_0)$ and $\mu_N(t_0)$; $\mu_N(t_0) = \mathcal{T}^0\#\rho(t_0)$. It is known that such an optimal transport map exists when $\rho(t_0)$ is absolutely continuous with respect to the Lebesgue measure Champion, Pascale, and Juutinen (2008). Then it follows from Theorem 4.1 that $\rho(t) = \Psi(t;t_0,\cdot\,)\#\rho(t_0)$ and obviously $\mu_N(t) = \Psi_N(t;t_0,\cdot\,)\#\mu_N(t_0)$ for $t > t_0$. We also notice that for $t \ge t_0$

$$
\mathcal{T}^t\#\rho(t) = \mu_N(t), \quad \text{where} \quad \mathcal{T}^t = \Psi_N(t;t_0,\cdot)\circ\mathcal{T}^0\circ\Psi(t_0;t,\cdot).
$$

By Definition 2.1 of the d_p Wasserstein distance, we get

$$
d_p^p(\mu_N(t),\rho(t)) \le \int_{\mathbb{R}^d} |\Psi(t;t_0,x) - \Psi_N(t;t_0,\mathcal{T}^0(x))|^p \rho(t_0,x)dx.
$$

In the case of $p = \infty$, we obtain

$$
\eta(t) = d_\infty(\mu_N(t),\rho(t)) \le \|\Psi(t;t_0,\cdot) - \Psi_N(t;t_0,\cdot)\circ\mathcal{T}^0\|_\infty.
$$

We notice that

$$\frac{d}{dt}\left(\Psi_N(t;t_0,\mathcal{T}^0(x)) - \Psi(t;t_0,x)\right)\Big|_{t=t_0} = u_N(t_0,\mathcal{T}^0(x)) - u(t_0,x).$$

Thus, writing the integral form, dividing by $t-t_0$, and taking the limit $t \to t_0^+$ we easily get

$$\frac{d}{dt}\|\Psi_N(t;t_0,\cdot)\circ\mathcal{T}^0 - \Psi(t;t_0,\cdot)\|_\infty\Big|_{t=t_0^+} \le \|u_N(t_0,\cdot)\circ\mathcal{T}^0 - u(t_0,\cdot)\|_\infty. \tag{3.28}$$

We now note that

$$\begin{aligned} u_N(t_0,\mathcal{T}^0(x)) &- u(t_0,x) \\ &= -\int_{\mathbb{R}^d} \nabla W(\mathcal{T}^0(x)-y)d\mu_N(t_0,y) + \int_{\mathbb{R}^d} \nabla W(x-y)\rho(t_0,y)dy \\ &= -\int_{\mathbb{R}^d} \left(\nabla W(\mathcal{T}^0(x)-\mathcal{T}^0(y)) - \nabla W(x-y)\right)\rho(t_0,y)dy. \end{aligned}$$

For notational simplicity, we omit the time dependency on t_0 in the next few computations. This yields that (3.28) can be rewritten as

$$\frac{d^+\eta}{dt} \le C \sup_{x\in\mathbb{R}^d} \int_{\mathbb{R}^d} |\nabla W(\mathcal{T}(x)-\mathcal{T}(y)) - \nabla W(x-y)|\rho(y)dy. \tag{3.29}$$

We decompose the integral on $\mathbb{R}^d$ into the near- and the far-field parts as $\mathcal{A} := \{z : |x-z| \ge 4\eta\}$ and $\mathcal{B} := \mathbb{R}^d - \mathcal{A}$ as

$$\begin{aligned} \int_{\mathbb{R}^d} |\nabla W(\mathcal{T}(x)-\mathcal{T}(y)) - \nabla W(x-y)|\rho(y)dy &= \int_{\mathcal{A}} \cdots + \int_{\mathcal{B}} \cdots \\ &:= \mathcal{I}_1 + \mathcal{I}_2. \end{aligned} \tag{3.30}$$

For the estimate in the set $\mathcal{A}$, we use

$$|\mathcal{T}(x)-\mathcal{T}(y)| \ge |x-y| - |\mathcal{T}(x)-x| - |\mathcal{T}(y)-y| \ge |x-y| - 2\eta \ge \frac{|x-y|}{2}$$

together with (3.26) and (3.27) to obtain

$$\begin{aligned} \mathcal{I}_1 &\le \int_{\mathcal{A}} \frac{2\left(|x-\mathcal{T}(x)| + |y-\mathcal{T}(y)|\right)}{\min(|x-y|,|\mathcal{T}(x)-\mathcal{T}(y)|)^{\alpha+1}}\rho(y)dy \\ &\le 4\eta\int_{\mathcal{A}}\left(\frac{1}{|x-y|^{\alpha+1}} + \frac{2^{\alpha+1}}{|x-y|^{\alpha+1}}\right)\rho(y)dy \le C\eta\int_{\mathcal{A}}\frac{1}{|x-y|^{\alpha+1}}\rho(y)dy \\ &\le C\eta\int_{\mathbb{R}^d}\frac{1}{|x-y|^{\alpha+1}}\rho(y)dy \le C\eta\|\rho\|. \end{aligned} \tag{3.31}$$

For the second part $\mathcal{I}_2$, we estimate separately each term using (3.20) to deduce

$$\begin{aligned}\mathcal{I}_2 &\leq \int_{\mathcal{B}} \frac{\rho(y)}{|x-y|^\alpha} dy + \int_{\mathcal{B}} \frac{\rho(y)}{\eta_m^\alpha} dy \\ &\leq \left(\int_{\mathcal{B}} \frac{1}{|x-y|^{\alpha p'}} dy \right)^{1/p'} \|\rho\|_p + \frac{1}{\eta_m^\alpha} \left(\int_{\mathcal{B}} 1 dy \right)^{1/p'} \|\rho\|_p \\ &\leq C(\eta^{d/p'-\alpha} + \eta^{d/p'} \eta_m^{-\alpha}) \|\rho\|_p \leq C(\eta^{d/p'-\alpha} + \eta^{d/p'} \eta_m^{-\alpha}) \|\rho\| \,.\end{aligned} \tag{3.32}$$

Notice that $|\mathcal{T}(x) - \mathcal{T}(y)| \geq \eta_m$ by definition of the minimum inter-particle distance (3.19) as soon as $\mathcal{T}(x) \neq \mathcal{T}(y)$, $\nabla W(\mathcal{T}(x) - \mathcal{T}(y)) = 0$ otherwise.

Finally, we choose two indices i, j so that $|X_i - X_j| = \eta_m$, then we observe that the middle point between X_i and X_j has to be transported by $\mathcal{T}$ to either X_i or X_j, and thus $\eta_m \leq 2\eta$. Hence by combining (3.29)-(3.32) and being t_0 arbitrary in $[0, \min(T, T_0^N))$, we have

$$\frac{d^+\eta}{dt} \leq C\eta \|\rho\| \left(1 + \eta^{d/p'-1} \eta_m^{-\alpha}\right) \leq C\eta \|\rho\| \left(1 + \eta^{d/p'} \eta_m^{-(1+\alpha)}\right) , \tag{3.33}$$

for all $t \in [0, \min(T, T_0^N))$.

Step B.- We now focus on showing the lower bound estimate of η_m to make the system (3.33) closed. We again choose two indices i, j so that $|X_i - X_j| = \eta_m$. Neglecting the time dependency to simplify the notation, we get

$$\begin{aligned}\frac{d}{dt}|X_i - X_j| &\geq -|u_N(X_i) - u_N(X_j)| \\ &\geq - \int_{\mathbb{R}^d} |\nabla W(X_i - y) - \nabla W(X_j - y)| \, d\mu_N(y) \\ &= - \int_{\mathbb{R}^d} |\nabla W(X_i - \mathcal{T}(y)) - \nabla W(X_j - \mathcal{T}(y))| \, \rho(y) dy \,,\end{aligned}$$

where $\mathcal{T}$ is the optimal map satisfying $\mu_N(t) = \mathcal{T}\#\rho(t)$, for each $t \in [0, \min(T, T_0^N))$. Similar to (3.30), we split in near- and far-field parts the domain $\mathbb{R}^d$ as $\mathcal{A} := \{y : |X_i - y| \geq 2\eta \text{ and } |X_j - y| \geq 2\eta\}$ and $\mathcal{B} := \mathbb{R}^d - \mathcal{A}$.

We can again use (3.26) to deduce

$$\int_{\mathcal{A}} |\nabla W(X_i - \mathcal{T}(y)) - \nabla W(X_j - \mathcal{T}(y))| \, \rho(y) dy \tag{3.34}$$
$$\leq \int_{\mathcal{A}} \frac{2|X_i - X_j|}{\min(|X_i - \mathcal{T}(y)|, |X_j - \mathcal{T}(y)|)^{\alpha+1}} \rho(y) dy$$
$$\leq 2^{2+\alpha} |X_i - X_j| \int_{\mathcal{A}} \left(\frac{1}{|X_i - y|^{\alpha+1}} + \frac{1}{|X_j - y|^{\alpha+1}} \right) \rho(y) dy \leq C \eta_m \|\rho\|,$$

where we used that $|X_i - \mathcal{T}(y)| \geq |X_i - y| - \eta \geq \frac{1}{2}|X_i - y|$ and similarly for X_j together with (3.27). For the integral over $\mathcal{B}$, we use that as soon as $X_i \neq \mathcal{T}(y)$, then we obtain from (3.20) that

$$|\nabla W(X_i - \mathcal{T}(y))| \leq \frac{1}{|X_j - \mathcal{T}(y)|^\alpha} \leq \frac{1}{\eta_m^\alpha},$$

and $\nabla W(X_i - \mathcal{T}(y)) = 0$ otherwise, and similarly for X_j. A simple Hölder computation as in (3.27) implies that

$$\int_{\mathcal{B}} \rho(y) dy \leq C \eta^{d/p'} \|\rho\|,$$

from which we infer that

$$\int_{\mathcal{B}} |\nabla W(X_i - \mathcal{T}(y)) - \nabla W(X_j - \mathcal{T}(y))| \, \rho(y) dy \leq C \eta^{d/p'} \eta_m^{-\alpha} \|\rho\|. \tag{3.35}$$

Putting together (3.34) and (3.35), we finally conclude that

$$\frac{d\eta_m}{dt} \geq -C \eta_m \|\rho\| \left(1 + \eta^{d/p'} \eta_m^{-(1+\alpha)}\right), \tag{3.36}$$

for all $t \in [0, \min(T, T_0^N))$.

Step C.- Until now, we have proved from (3.33) and (3.36) that

$$\begin{cases} \dfrac{d^+\eta}{dt} & \leq C\eta \|\rho\| \left(1 + \eta^{d/p'} \eta_m^{-(1+\alpha)}\right), \\ \dfrac{d\eta_m}{dt} & \geq -C \eta_m \|\rho\| \left(1 + \eta^{d/p'} \eta_m^{-(1+\alpha)}\right), \end{cases} \tag{3.37}$$

for $t \in [0, \min(T, T_0^N))$. We first notice from (3.37) that if $\eta^{d/p'} \eta_m^{-(1+\alpha)} \leq 1$, then

$$\eta(t) \leq \eta^0 e^{2\|\rho\| t} \quad \text{and} \quad \eta_m(t) \geq \eta_m^0 e^{-2\|\rho\| t} \quad t \in [0, \min(T, T_0^N)). \tag{3.38}$$

We now show that (3.38) holds for time $t \in [0,T]$ when N goes to infinity, in other words that $T < T_0^N$ when N is sufficiently large. For this, we set

$$f(t) := \frac{\eta(t)}{\eta^0}, \quad g(t) := \frac{\eta_m(t)}{\eta_m^0} \quad \text{and} \quad \xi_N := (\eta^0)^{d/p'}(\eta_m^0)^{-(1+\alpha)}.$$

Note that ξ_N depends on the number of particles N as in (3.19). It yields

$$\begin{aligned} \frac{d^+ f}{dt} &\leq C\|\rho\|\, f\left(1+\xi_N f^{d/p'} g^{-(1+\alpha)}\right), \\ \frac{dg}{dt} &\geq -C\|\rho\|\, g\left(1+\xi_N f^{d/p'} g^{-(1+\alpha)}\right). \end{aligned}$$

Since $f(0) = g(0) = 1$ and $\xi_N \to 0$ as N goes to infinity, we obtain that there exists a positive constant $T_*^N (\leq T_0^N)$ such that

$$\xi_N f^{d/p'} g^{-(1+\alpha)} \leq 1 \quad \text{for} \quad t \in [0, T_*^N],$$

for sufficiently large N. Then it follows from (3.38) that

$$f(t) \leq e^{2\|\rho\| t} \quad \text{and} \quad g(t) \geq e^{-2\|\rho\| t}.$$

This yields $\xi_N f^{d/p'} g^{-(1+\alpha)} \leq \xi_N e^{2(d/p'+(1+\alpha))\|\rho\| t}$, that is,

$$\xi_N f^{d/p'} g^{-(1+\alpha)} \leq 1 \quad \text{holds for} \quad t \leq -\frac{\ln(\xi_N)}{2(d/p' + (1+\alpha))\|\rho\|},$$

so that

$$-\frac{\ln(\xi_N)}{2(d/p' + (1+\alpha))\|\rho\|} \leq T_*^N.$$

On the other hand, our assumption for the initial data (3.21) implies

$$\liminf_{N\to\infty} T_*^N \geq \lim_{N\to\infty} -\frac{\ln(\xi_N)}{2(d/p' + (1+\alpha))\|\rho\|} = \infty,$$

and thus for N large enough, $T < T_*^N < T_0^N$. This completes the proof. □

Remark 3.3. One can use almost the same argument with the above to obtain an stability estimate in d_∞: let ρ_1 and ρ_2 be solutions given by Theorem 4.1 to the system (3.16) satisfying (3.20), then we have

$$\frac{d}{dt} d_\infty(\rho_1(t), \rho_2(t)) \leq C \max(\|\rho_1\|, \|\rho_2\|)\, d_\infty(\rho_1(t), \rho_2(t)).$$

In fact, the estimate of mean field limit in Theorem 3.1 holds for $-1 \leq \alpha < 0$ without any condition on η^0 and η^0_m. This is coherent with the results in Carrillo, Di Francesco, Figalli, Laurent, and Slepčev (2011) in which the mean field limit is obtained for all measure initial data without restriction in the way initial data are approximated by Dirac masses at least for attractive potentials.

Corollary 3.4. *Suppose the interaction potential W satisfies* (3.20) *with $-1 \leq \alpha < 0$, and let ρ be a solution to the system* (3.16) *such that $\rho \in L^\infty(0,T;(L^1 \cap L^p)(\mathbb{R}^d)) \cap \mathcal{C}([0,T],\mathcal{P}_1(\mathbb{R}^d))$. Suppose that*

$$d_\infty(\mu_N^0, \rho^0) \to 0 \quad as \quad N \to \infty.$$

Then for any solution of the ODE system (3.17) *the associated empirical distributions $\mu_N(t)$ converge toward $\rho(t)$ uniformly in time:*

$$\sup_{t\in[0,T]} d_\infty(\mu_N(t), \rho(t)) \to 0 \quad as \quad N \to \infty.$$

Remark 3.5. It is remarkable that even if we do not have uniqueness of solution of (3.17) under assumption (3.20) with $-1 \leq \alpha < 0$, we get the mean field limit without restriction. If one collision occurs, then uniqueness may lost, but the existence of solution is still guaranteed. Thus Corollary 3.4 is interesting because it is valid for density solutions to (3.16) even if collisions occur and uniqueness is lost at the particle level.

Proof of Corollary 3.4. We first notice that the existence of solutions to the ODE system (3.17) is guaranteed thanks to Cauchy-Peano-Arzela theorem since α is strictly negative with (3.20) implies that ∇W is continuous. One can use the same arguments as in the proof of Theorem 3.1 to find

$$\begin{aligned}\frac{d^+\eta}{dt} &\leq C \sup_{x\in\mathbb{R}^d}\left(\int_{\mathcal{A}} + \int_{\mathcal{B}}\right) |\nabla W(\mathcal{T}(x) - \mathcal{T}(y)) - \nabla W(x-y)|\, \rho(y)dy \\ &:= \mathcal{K}_1 + \mathcal{K}_2,\end{aligned}$$

where the same notation for the sets $\mathcal{A}$ and $\mathcal{B}$ is used and the time dependency has been avoided for simplicity. Using (3.31) we estimate $\mathcal{K}_1$ by $C\eta\|\rho\|$. To estimate $\mathcal{K}_2$, we use that $\alpha < 0$ to get

$$|\nabla W(\mathcal{T}(x) - \mathcal{T}(y)) - \nabla W(x-y)| \leq \frac{C}{\eta^\alpha} + \frac{C}{|x-y|^\alpha},$$

and to obtain by Hölder's inequality that

$$\begin{aligned}\mathcal{K}_2 &\leq C\int_{\mathcal{B}} \frac{\rho(y)}{|x-y|^\alpha} dy + \frac{C}{\eta^\alpha}\int_{\mathcal{B}} \rho(y)dy \leq C\eta^{d/p'-\alpha}\|\rho\|_p + C\eta^{d/p'}\eta^{-\alpha}\|\rho\|_p \\ &\leq C\eta^{d/p'-\alpha}\|\rho\|\,.\end{aligned}$$

Hence, we have

$$\frac{d^+\eta}{dt} \le C\eta\|\rho\| \left(1 + \eta^{d/p'-\alpha-1}\right),$$

and this yields for sufficiently large N

$$\eta(t) \le \left((\eta^0)^{1-(d/p'-\alpha)} e^{-C\|\rho\|(d/p'-\alpha-1)t} + e^{-C\|\rho\|(d/p'-\alpha-1)t} - 1 \right)^{-\frac{1}{d/p'-\alpha-1}},$$

for all $t \in [0, T]$. Note that $d/p' - \alpha - 1 > 0$ and then, the right hand side of previous estimate goes to zero as N goes to infinity. This completes the proof. □

We next show that there is no collision between particles when the initial quantities η^0 and η_m^0 in (3.19) satisfy

$$\lim_{N\to\infty} \frac{(\eta^0)^{d/p'-\alpha}}{\eta_m^0} = 0. \tag{3.39}$$

Note that the same strategy as in Remark 3.2 allows us to find suitable approximations for the initial data satisfying (3.39).

Corollary 3.6. *Under the assumptions of Corollary 3.4 with $-1 \le \alpha < 0$, if we further assume that η^0, η_m^0 satisfy (3.39). Then we have that for N large enough, the particle system (3.17) is uniquely well-defined till time T in the sense that there is no collision between particles before that time, and the convergence*

$$\sup_{t\in[0,T]} d_\infty(\mu_N(t), \rho(t)) \to 0 \quad as \quad N \to \infty,$$

holds.

Proof. The proof of Corollary 3.4 shows that for sufficiently large N

$$\eta \le \left((\eta^0)^{1-(d/p'-\alpha)} e^{-C\|\rho\|(d/p'-\alpha-1)t} + e^{-C\|\rho\|(d/p'-\alpha-1)t} - 1 \right)^{-\frac{1}{d/p'-\alpha-1}}.$$

For the estimate of η_m, one can obtain from the proof of Theorem 3.1 that

$$\frac{d\eta_m}{dt} \ge -C\eta_m\|\rho\| \left(1 + \eta^{d/p'-\alpha}\eta_m^{-1}\right) \quad \text{for all} \quad t \in [0, \min(T, T_0^N)),$$

where T_0^N denotes the first collision time between particles. Then we conclude the desired result employing the same arguments in Step C of Theorem 3.1 using (3.39). □

As a corollary of Theorem 3.1, we consider interaction potentials under weaker assumptions than (3.20): there exists $R > 0$ such that W satisfies

$$|\nabla W(x)| \leq \frac{C}{|x|^\alpha}, \quad \text{and} \quad |D^2 W(x)| \leq \frac{C}{|x|^{1+\alpha}}, \quad \forall\, x \in B(0,R), \qquad (3.40)$$

where $B(0,R) := \{x \in \mathbb{R}^d : |x| < R\}$. Then one can assume that the initial data ρ^0 has compact support, and show that the local solution $\rho(t)$ has compact support on a small time interval $[0,T]$. This is possible since characteristics are locally in time well defined and the velocity is uniformly bounded under the assumptions (3.40) initially. This argument was made rigorous under stricter assumptions on the local behaviour of the interaction potential but allowing growth of the potential at infinity in Balagué and Carrillo (2012). Thus, one can cut-off the potential outside a large ball in such a way that the solution is unaffected but the potential satisfies the global assumption $\nabla W \in \mathcal{W}^{1,p'}(\mathbb{R}^d)$ entering the well-posedness theory in Bertozzi, Laurent, and Rosado (2010) or satisfying (3.20) allowing for the application of Theorem 4.1. Concerning the interaction potential W satisfying (3.40), the same results of convergence in Theorem 3.1 and Corollary 3.6 can be obtained. We leave the details to the reader.

4 Local existence and uniqueness of L^p-solutions

In this section, we provide a local existence and uniqueness result of weak solutions in L^p-spaces to the system (3.16) under the assumptions (3.20).

As we mentioned before, we can not directly apply the arguments in Bertozzi, Laurent, and Rosado (2010) for the potentials satisfying (3.20). Of course, we can overcome these difficulties using the property of compact supports on the initial data ρ^0 (see the paragraph below Corollary 3.6). However, we use the arguments of dividing near- and far-field parts of the interaction potential function W to establish the local existence of a unique L^p-solution to the continuity aggregation equation (3.16).

Theorem 4.1. *Assume that W satisfies the condition (3.20), for some $-1 < \alpha < \frac{d}{p'} - 1$, and that $\rho^0 \in \mathcal{P}_1(\mathbb{R}^d) \cap L^p(\mathbb{R}^d)$, $1 < p \leq \infty$. Then there exists a time $T > 0$, depending only on $\|\rho^0\|_p$ and α, and a unique nonnegative solution to (3.16) satisfying $\rho \in L^\infty(0,T; L^1 \cap L^p(\mathbb{R}^d)) \cap \mathcal{C}([0,T], \mathcal{P}_1(\mathbb{R}^d))$. Furthermore, the solution satisfies that there exists $C > 0$ depending only on $\|\rho^0\|_p$ and α such that*

$$\|\rho(t)\|_p \leq C \qquad \textit{for all } t \in [0,T]. \qquad (4.41)$$

The velocity field generated by ρ, given by $u = -\nabla W * \rho$, is bounded and Lipschitz continuous in space uniformly on $[0,T]$, and ρ is determined as the push-forward of the initial density through the flow map generated by u.

Moreover, if ρ_i, $i = 1,2$, are two such solutions to (3.16) with initial conditions $\rho_i^0 \in \mathcal{P}_1(\mathbb{R}^d) \cap L^p(\mathbb{R}^d)$, $1 < p \leq \infty$, we have the following stability estimate:

$$\frac{d}{dt} d_1(t) \leq C \max(\|\rho_1\|, \|\rho_2\|) d_1(t),$$

where $d_1(t) := d_1(\rho_1(t), \rho_2(t))$.

Proof. Let us start by proving the uniqueness. Given two weak solutions $\rho_i \in L^\infty(0,T; L^1 \cap L^p(\mathbb{R}^d)) \cap \mathcal{C}([0,T], \mathcal{P}_1(\mathbb{R}^d))$, $i = 1,2$, to the continuous aggregation equations (3.16), consider the two flow maps $\Psi_i : \mathbb{R}_+ \times \mathbb{R}_+ \times \mathbb{R}^d \to \mathbb{R}^d$, $i = 1,2$, generated by the two velocity fields, i.e.,

$$\begin{cases} \dfrac{d}{dt}\left(\Psi_i(t;s,x)\right) = u_i(t;s,\Psi_i(t;s,x))\,, \\ \Psi_i(s;s,x) = x, \end{cases}$$

where $u_i := -\nabla W * \rho_i$, $t, s \in [0,T]$ and $x \in \mathbb{R}^d$. We know that the solutions are constructed by transporting the initial measures through the velocity fields $\rho_i = \Psi_i \# \rho_i^0$, $i = 1,2$.

Let $\mathcal{T}^0$ be the optimal transportation between $\rho_1(0)$ and $\rho_2(0)$ for the d_1-distance. Then we define a transport (not necessarily optimal) between $\rho_1(t)$ and $\rho_2(t)$ by

$$\mathcal{T}^t(x) = \Psi_2(t;0,x) \circ \mathcal{T}^0(x) \circ \Psi_1(0;t,x), \qquad \mathcal{T}^t \# \rho_1(t) = \rho_2(t),$$

and $\frac{d}{dt} d_1(t) \leq Q(t)$, where $d_1(t) := d_1(\rho_1(t), \rho_2(t))$ and

$$Q(t) := \int_{\mathbb{R}^d \times \mathbb{R}^d} |\nabla W(\mathcal{T}^t(x) - \mathcal{T}^t(y)) - \nabla W(x-y)| \rho_1(t,x) \rho_1(t,y)\, dx dy\,,$$

where we have used a similar argument as in Step A of the proof of Theorem 3.1. To simplify the notation, let us not make explicit the dependence on time. Note by symmetry that

$$\begin{aligned} Q(t) &\leq 4 \int_{\mathbb{R}^d \times \mathbb{R}^d} \left(\frac{|\mathcal{T}(x) - x|}{|\mathcal{T}(x) - \mathcal{T}(y)|^{1+\alpha}} + \frac{|\mathcal{T}(x) - x|}{|x-y|^{1+\alpha}} \right) \rho_1(x) \rho_1(y)\, dx dy \\ &:= \mathcal{J}_1 + \mathcal{J}_2. \end{aligned}$$

Straightforward computation using the near- and far-field decomposition as in (3.27) shows that

$$\begin{aligned}\mathcal{J}_1 &= 4\int_{\mathbb{R}^d} |\mathcal{T}(x) - x|\rho_1(x)\left(\int_{\mathbb{R}^d} \frac{\rho_2(y)}{|\mathcal{T}(x)-y|^{1+\alpha}}dy\right)dx \\ &\leq C\|\rho_2\| \int_{\mathbb{R}^d} |\mathcal{T}(x) - x|\rho_1(x)dx = C\|\rho_2\|\, d_1(t).\end{aligned}$$

Similarly using again (3.27), we have $\mathcal{J}_2 \leq C\|\rho_1\| d_1(t)$. It yields that

$$\frac{d}{dt}d_1(t) \leq C\max(\|\rho_1\|, \|\rho_2\|)\, d_1(t)\,,$$

from which we conclude the uniqueness part of the statement.

Let us now show the existence of weak solution. Let $\varepsilon > 0$ and θ be a standard mollifier:

$$\theta \geq 0, \quad \theta \in \mathcal{C}_0^\infty(\mathbb{R}^d), \quad \text{supp}\, \theta \subset B(0,1), \quad \int_{\mathbb{R}^d} \theta(x)dx = 1,$$

and we set a sequence of smooth mollifiers:

$$\theta_\varepsilon(x) := \frac{1}{\varepsilon^d}\theta\left(\frac{x}{\varepsilon}\right).$$

We first regularize ∇W such as $\nabla W_\varepsilon := (\nabla W) * \theta_\varepsilon$. Then since ∇W_ε is a globally Lipschitz, we can apply the theory of Braun and Hepp (1977); Dobrushin (1979); Laurent (2007) which says that there exists a unique global solution ρ_ε to the following system

$$\begin{cases} \partial_t \rho_\varepsilon + \nabla\cdot(\rho_\varepsilon u_\varepsilon) = 0, & t > 0, \quad x \in \mathbb{R}^d, \\ u_\varepsilon(t,x) := -\nabla W_\varepsilon * \rho_\varepsilon, & t > 0, \quad x \in \mathbb{R}^d, \\ \rho_\varepsilon(0,x) := \rho^0(x), & x \in \mathbb{R}^d, \end{cases} \tag{4.42}$$

A standard calculation, see Bertozzi, Laurent, and Rosado (2010), implies that

$$\frac{d}{dt}\|\rho_\varepsilon\|_{L^1\cap L^p} \leq C\|\rho_\varepsilon\|^2_{L^1\cap L^p}, \tag{4.43}$$

where C is an uniform constant in ε. Note that the inequality (4.43) holds only formally for the non regularized problem, but it is fully rigorous for the regularized one with W_ε. This yields that the time of blow-up depends only on the initial data, more precisely $\|\rho^0\|$, and not on ε. Thus, there exists a $T > 0$ such that

$$\sup_{\varepsilon>0} \|\rho_\varepsilon\| < \infty. \tag{4.44}$$

It follows from (4.44) and the evolution in time of the first momentum of ρ, that this first moment is also uniformly bounded:

$$\sup_{\varepsilon>0} \|x\rho_\varepsilon\|_{L^\infty(0,T;L^1(\mathbb{R}^d))} \le C,$$

where C depends only on $T, \|x\rho^0\|_1$, and $\|\rho^0\|$. We leave the details to the reader. Next, we show an estimate on the growth of the d_1 distance $\eta_{\varepsilon,\varepsilon'}(t) := d_1(\rho_\varepsilon(t), \rho_{\varepsilon'}(t))$ between ρ_ε and $\rho_{\varepsilon'}$, for $\varepsilon, \varepsilon' > 0$:

$$\frac{d}{dt}\eta_{\varepsilon,\varepsilon'}(t) \le C \max(\|\rho_\varepsilon\|, \|\rho_{\varepsilon'}\|) \left(\eta_{\varepsilon,\varepsilon'}(t) + \varepsilon + \varepsilon'\right), \tag{4.45}$$

where C is an uniform constant in ε and ε'. We remark that the above estimate (4.45) implies that $\{\rho_\varepsilon\}_{\varepsilon>0}$ is a Cauchy sequence in $\mathcal{C}([0,T], \mathcal{P}_1(\mathbb{R}^d))$.

Let us remark that the weak solutions to the regularized problems (4.42) can be written in terms of characteristics. This is a consequence of the fact that the associated velocity field u_ϵ is bounded and Lipschitz in space, unifromly in time and some standard duality arguments. This strategy is explained in detail at the end of the proof of the present Theorem applied to the solution of the original problem, and we refer the reader there for details. Since solutions are constructed by characteristics, for the proof of (4.45) we can proceed as in the part of uniqueness. Therefore, not making explicit the time dependency, we get

$$\begin{aligned}
\frac{d}{dt}\eta_{\varepsilon,\varepsilon'}(t) &\le \int_{\mathbb{R}^d\times\mathbb{R}^d} |\nabla W_\varepsilon(\mathcal{T}(x) - \mathcal{T}(y)) - \nabla W_{\varepsilon'}(x-y)|\, \rho_{\varepsilon'}(x)\rho_{\varepsilon'}(y)dxdy \\
&\le \int_{\mathbb{R}^d\times\mathbb{R}^d} |\nabla W_\varepsilon(\mathcal{T}(x) - \mathcal{T}(y)) - \nabla W_\varepsilon(x-y)|\, \rho_{\varepsilon'}(x)\rho_{\varepsilon'}(y)dxdy \\
&\quad + \int_{\mathbb{R}^d\times\mathbb{R}^d} |\nabla W_\varepsilon(x-y) - \nabla W_{\varepsilon'}(x-y)|\, \rho_{\varepsilon'}(x)\rho_{\varepsilon'}(y)dxdy \\
&:= \mathcal{K}_1 + \mathcal{K}_2,
\end{aligned} \tag{4.46}$$

where $\mathcal{T}$ is the optimal transportation between $\rho_{\varepsilon'}(t)$ and $\rho_\varepsilon(t)$ for the d_1-distance. To estimate $\mathcal{K}_1$, we notice that

$$\begin{aligned}
|\nabla W_\varepsilon(x)| &\le \int_{\{y:|y|<\frac{|x|}{2}\}} \frac{\theta_\varepsilon(y)}{|x-y|^{1+\alpha}}dy + \int_{\{y:|y|\ge\frac{|x|}{2}\}} \frac{\theta_\varepsilon(y)}{|x-y|^{1+\alpha}}dy \\
&\le \frac{2^{1+\alpha}}{|x|^{1+\alpha}} \int_{\mathbb{R}^d} \theta_\varepsilon(y)dy + \mathbf{1}_{\{|x|\le 2\varepsilon\}} \int_{\{y:\ \varepsilon\ge|y|\}} \frac{\theta_\varepsilon(y)}{|x-y|^{1+\alpha}}dy \\
&\le \frac{C}{|x|^{1+\alpha}} + \frac{C\varepsilon^{1+\alpha}}{|x|^{1+\alpha}} \int_{\{y:\ \varepsilon\ge|y|\}} \frac{\theta_\varepsilon(y)}{|x-y|^{1+\alpha}}dy \le \frac{C}{|x|^{1+\alpha}}.
\end{aligned} \tag{4.47}$$

Then we now use again the decomposition (3.27) as in the part of uniqueness to find

$$\mathcal{K}_1 \le C \max(\|\rho_\varepsilon\|, \|\rho_{\varepsilon'}\|)\, \eta_{\varepsilon,\varepsilon'}(t)\,, \tag{4.48}$$

where C, $\|\rho_\varepsilon\|$, and $\|\rho_{\varepsilon'}\|$ are uniformly bounded in ε and ε' thanks to the estimate (4.44). For the estimate of $\mathcal{K}_2$, we claim that

$$|\nabla(W - W_\varepsilon)(x)| \le \frac{C\varepsilon}{|x|^{1+\alpha}}, \tag{4.49}$$

where C is independent on ε.

Proof of Claim: It is a straightforward to obtain

$$\begin{aligned}|\nabla W_\varepsilon(x) - \nabla W(x)| &\le \int_{\mathbb{R}^d} |\nabla W(x-y) - \nabla W(x)|\theta_\varepsilon(y)dy \\ &\le 2\int_{\mathbb{R}^d}\left(\frac{1}{|x|^{1+\alpha}} + \frac{1}{|x-y|^{1+\alpha}}\right)|y|\theta_\varepsilon(y)dy \\ &:= \mathcal{L}_1 + \mathcal{L}_2\,.\end{aligned} \tag{4.50}$$

Noticing that the mollifier properties allow to gain an ε factor in front of the integrals, we can estimate $\mathcal{L}_i, i = 1, 2$ as follows

$$\begin{aligned}\mathcal{L}_1 &\le \frac{C\varepsilon}{|x|^{1+\alpha}}\int_{\mathbb{R}^d}\theta_\varepsilon(y)dy = \frac{C\varepsilon}{|x|^{1+\alpha}}, \\ \mathcal{L}_2 &\le 2\varepsilon\int_{\mathbb{R}^d}\frac{\theta_\varepsilon(y)}{|x-y|^{1+\alpha}}dy \le \frac{C\varepsilon}{|x|^{1+\alpha}},\end{aligned} \tag{4.51}$$

where we used a similar argument to (4.47) for $\mathcal{L}_2$. We now combine (4.50) and (4.51) to have the inequality (4.49). Then we use (4.49) together with (3.27) to find the estimate of $\mathcal{K}_2$

$$\mathcal{K}_2 \le C(\varepsilon + \varepsilon')\int_{\mathbb{R}^d\times\mathbb{R}^d}\frac{\rho_{\varepsilon'}(t,x)\rho_{\varepsilon'}(t,y)}{|x-y|^{1+\alpha}}dxdy \le C(\varepsilon+\varepsilon')\|\rho_{\varepsilon'}\|\,. \tag{4.52}$$

This completes the proof of the inequality (4.45) by combining (4.45), (4.46), (4.48), and (4.52).

Since ρ_ε is a Cauchy sequence in $\mathcal{C}([0,T],\mathcal{P}_1(\mathbb{R}^d))$, it converges toward a limit curve of measures $\rho \in \mathcal{C}([0,T],\mathcal{P}_1(\mathbb{R}^d))$, and we also have $\rho \in L^\infty(0,T;L^1\cap L^p(\mathbb{R}^d))$ from the uniform bounded estimate (4.44). It remains to show that ρ is a solution of the aggregation equations (3.16). Choose a test function $\phi(t,x) \in \mathcal{C}_c^\infty([0,T]\times\mathbb{R}^d)$, then ρ_ε satisfies

$$\begin{aligned}\int_{\mathbb{R}^d}\rho^0(x)\phi^0(x)dx &= \int_{\mathbb{R}^d}\rho_\varepsilon(T,x)\phi(T,x)dx + \int_0^T\int_{\mathbb{R}^d}\rho_\varepsilon(t,x)\partial_t\phi(t,x)dxdt \\ &\quad - \int_0^T\int_{\mathbb{R}^d}\int_{\mathbb{R}^d}\rho_\varepsilon(t,x)\rho_\varepsilon(t,y)\nabla W_\varepsilon(x-y)\cdot\nabla\phi(t,x)dxdydt.\end{aligned} \tag{4.53}$$

The first two terms in the rhs of (4.53) converges to

$$\int_{\mathbb{R}^d} \rho(T,x)\phi(T,x)dx + \int_0^T \int_{\mathbb{R}^d} \rho(t,x)\partial_t\phi(t,x)dxdt,$$

since $\rho_\varepsilon \to \rho$ in $\mathcal{C}([0,T],\mathcal{P}_1(\mathbb{R}^d))$. For the third term in the rhs of (4.53), we use the estimates (4.49) and (4.44) to find

$$\left|\int_0^T\int_{\mathbb{R}^d}\int_{\mathbb{R}^d}\rho_\varepsilon(t,x)\rho_\varepsilon(t,y)\left(\nabla W_\varepsilon(x-y)-\nabla W(x-y)\right)\cdot\nabla\phi(t,x)dxdydt\right| \to 0,$$

as $\varepsilon \to 0$. It remains to show that

$$\int_0^T\int_{\mathbb{R}^d}\int_{\mathbb{R}^d}\rho_\varepsilon(t,x)\rho_\varepsilon(t,y)\nabla W(x-y)\cdot\nabla\phi(t,x)dxdydt$$
$$\to \int_0^T\int_{\mathbb{R}^d}\int_{\mathbb{R}^d}\rho(t,x)\rho(t,y)\nabla W(x-y)\cdot\nabla\phi(t,x)dxdydt,$$

as $\varepsilon \to 0$. For this, we introduce a cut-off function $\chi_\delta \in \mathcal{C}_c^\infty(\mathbb{R})$ such that

$$\chi_\delta(x) = \begin{cases} 1 & \text{if} \quad |x| \le \delta \\ 0 & \text{if} \quad |x| \ge 2\delta \end{cases}.$$

Then it follows from the weak convergence that

$$\int_0^T\int_{\mathbb{R}^d}\int_{\mathbb{R}^d}\rho_\varepsilon(t,x)\rho_\varepsilon(t,y)(1-\chi_\delta(x-y))\nabla W(x-y)\cdot\nabla\phi(t,x)dxdydt$$
$$\to \int_0^T\int_{\mathbb{R}^d}\int_{\mathbb{R}^d}\rho(t,x)\rho(t,y)(1-\chi_\delta(x-y))\nabla W(x-y)\cdot\nabla\phi(t,x)dxdydt,$$

as $\varepsilon \to 0$, since $(1-\chi_\delta(x-y))\nabla W(x-y)\cdot\nabla\phi(t,x)$ is a Lipschitz function. We estimate the remainder as follows:

$$\left|\int_0^T\int_{\mathbb{R}^d}\int_{\mathbb{R}^d}\rho_\varepsilon(t,x)\rho_\varepsilon(t,y)\chi_\delta(x-y))\nabla W(x-y)\cdot\nabla\phi(t,x)dxdydt\right|$$
$$\le C\delta\int_0^T\int_{\{(x,y)\in\mathbb{R}^d\times\mathbb{R}^d:\ |x-y|\le 2\delta\}}\frac{1}{|x-y|^{1+\alpha}}\rho_\varepsilon(t,x)\rho_\varepsilon(t,y)dxdydt$$
$$\le CT\delta\|\rho_\varepsilon\| \to 0 \quad \text{as} \quad \delta \to 0.$$

Similarly, we have

$$\lim_{\delta\to 0}\left|\int_0^T\int_{\mathbb{R}^d}\int_{\mathbb{R}^d}\rho(t,x)\rho(t,y)\chi_\delta(x-y)\nabla W(x-y)\cdot\nabla\phi(t,x)dxdydt\right| = 0.$$

Hence, we conclude that ρ satisfies

$$\int_{\mathbb{R}^d} \rho^0(x)\phi^0(x)dx = \int_{\mathbb{R}^d} \rho(T,x)\phi(T,x)dx + \int_0^T \int_{\mathbb{R}^d} \rho(t,x)\partial_t\phi(t,x)dxdt \quad (4.54)$$
$$- \int_0^T \int_{\mathbb{R}^d} \int_{\mathbb{R}^d} \rho(t,x)\rho(t,y)\nabla W(x-y)\cdot\nabla\phi(t,x)dxdydt,$$

for all $\phi \in \mathcal{C}_c^\infty([0,T]\times\mathbb{R}^d)$.

Now, We notice that a weak solution in $\rho \in L^\infty(0,T;L^1\cap L^p(\mathbb{R}^d))$ to (3.16) under the assumptions (3.20) has a well defined flow by using the same arguments as the ones at the beginning of Theorem 3.1. In fact, the velocity field is bounded and Lipschitz continuous in space with

$$|u(t,x)-u(t,y)| \le C\|\rho\||x-y|$$

for all $x,y\in\mathbb{R}^d$ and $t\in[0,T]$. Thus, the flow map

$$\begin{cases} \dfrac{d}{dt}(\Psi(t;s,x)) = u(t;s,\Psi(t;s,x)), \\ \Psi(s;s,x) = x, \end{cases}$$

for all $s,t \in [0,T]$ is well-defined. Choosing as test function in (4.54) $\phi(t,x)=\varphi(\Psi(t;\bar{T},x))$ for any $\bar{T}\in(0,T]$ with $\varphi\in\mathcal{C}_c^\infty(\mathbb{R}^d)$, it is a straightforward to check, due to the definition of the flow map, that

$$\int_{\mathbb{R}^d} \rho^0(x)\varphi(\Psi(0;\bar{T},x))dx = \int_{\mathbb{R}^d} \rho(\bar{T},x)\varphi(x)dx,$$

for all $\varphi\in\mathcal{C}_c^\infty(\mathbb{R}^d)$, and thus by a density argument we conclude $\rho(\bar{T}) = \Psi(\bar{T};0,\cdot\,)\#\rho^0$. Since this argument can be done for all $0<\bar{T}\le T$, this completes the proof. □

5 Propagation of chaos

In most practical purposes to approximate the continuum model by particle systems, it is naturally expected that initial positions and velocities will randomly and independently be selected. We will show that the empirical measure at time 0 is then close to ρ^0 with large probability in suitable weak norm.

In a seminal article Kac (1956), the propagation of chaos was introduced by Kac giving a proof for a simplified collision evolution process. He showed how the limit of many particles rigorously follows from the property of propagation of chaos. For a classical introduction to these topics, we refer to Sznitman (1991). Later, this property has been studied and developed in kinetic theory, Mckean (1967, 1975); Graham and Méléard (1997); Hauray and Mischler (2012); Mischler and Mouhot (2013).

Let us introduce the notion of propagation of chaos. Let us consider $\rho^N(t, x_1, \cdots, x_N)$ being the image by the dynamics to the coupled system (3.17) with N-equal masses particles of the initial law $(\rho^0)^{\otimes N}$. We define the k-marginals as follows.

$$\rho_k^N(t, x_1, \cdots, x_k) := \int_{\mathbb{R}^{d(N-k)}} \rho_N(t, x) dx_{k+1} \cdots, dx_N.$$

Let us choose the initial positions $X^{N,0} := \{X_i^0\}_{i=1}^N$ as independent identically distributed random variables (in short iid) with law ρ^0. We can construct the associated empirical measure as in (3.18) by

$$\mu_N(t) = \frac{1}{N} \sum_{i=1}^{N} \delta_{X_i(t)},$$

but now understood as a random variable with values in the space of probability measures.

The propagation of chaos property is defined as follows: for any fixed $k \in \mathbb{N}$,

$$\rho_k^N \rightharpoonup (\rho)^{\otimes k} \quad \text{weakly-}* \text{ as measures as} \quad N \to \infty.$$

It is classically known Sznitman (1991) that it is sufficient to check this property for $k = 2$ to derive the propagation of chaos. In fact, this is based on the fact that propagation of chaos is equivalent to show that the empirical measures $\mu_N(t)$ converge in law towards the constant random variable $\rho(t)$.

Theorem 5.1 gives a quantified version of the convergence in probability of $\mu_N(t)$ towards $\rho(t)$. We refer to Hauray and Mischler (2012); Mischler and Mouhot (2013) for a detailed explanation of the quantified equivalence relations. The propagation of chaos for the Vlasov-Poisson equations with singular force has recently been investigated in Hauray and Jabin (2012). Here, we are only able to provide such a result in a more restrictive setting that in the previous section. Namely, we only show the propagation of chaos for $d \geq 3$ and with a more restrictive condition on the allowed singularities $\alpha \geq 0$ depending on the regularity of the initial data $1 < p < \infty$.

Theorem 5.1. *Given* $\rho(t) \in L^\infty(0, T; (L^1 \cap L^p)(\mathbb{R}^d)) \cap \mathcal{C}([0, T], \mathcal{P}_1(\mathbb{R}^d))$ *the unique solution to* (3.16) *with initial data* $\rho^0 \in \mathcal{P}_1(\mathbb{R}^d) \cap L^p(\mathbb{R}^d)$, $1 < p \leq \infty$, *up to time $T > 0$. Assume that ρ^0 has compact support, that the initial positions $X^{N,0} := \{X_i^0\}_{i=1}^N$ are iid with law ρ^0, and that*

$$(1+\alpha)p' < \frac{p-1}{2p-1}d,$$

with $\alpha \geq 0$. Then the propagation of chaos holds in the sense that

$$\mathbb{P}\left(\sup_{t\in[0,T]} d_1(\mu_N(t),\rho(t)) \geq \frac{C}{N^{\gamma/d}}\right) \to 0, \quad \text{as } N \to +\infty,$$

where γ is a positive constant satisfying

$$\frac{p'(2p-1)(1+\alpha)}{d(p-1)} < \gamma < 1.$$

Remark 5.2. The condition on α gets more and more restrictive as p gets smaller and smaller. In $d = 2$, even for $p = \infty$ the condition is empty for $\alpha \geq 0$. In $d = 3$, you get the condition $\alpha < 1/2$ for $p = \infty$ and with $p = \frac{5+\sqrt{13}}{2}$ the condition is already empty. We also notice that the existence and uniqueness of the solutions are guaranteed by Theorem 4.1 and taking expectations in the corresponding inequalities for the particle system. Finally, in case $-1 \leq \alpha < 0$, the propagation of chaos holds using the same strategy as in Corollary 3.4 by taking expectations in the inequalities for the evolution of the Wasserstein distance.

We will follow the strategy recently introduced in Hauray and Jabin (2012) for the Vlasov equation. We first find a deterministic version of the propagation of chaos. This means that we consider a regularized system of particles as a kind of middle ground between the solution of the mean-field equation (3.16) and the random particle evolution. More precisely, we define the "blob" initial data ρ_N^0 as

$$\rho_N^0 := \mu_N^0 * \frac{\mathbf{1}_{B_\varepsilon(0)}}{|B_\varepsilon(0)|} = \frac{1}{c_d\varepsilon^d}\left(\mu_N^0 * \mathbf{1}_{B_\varepsilon(0)}\right), \tag{5.55}$$

where $\varepsilon > 0$ to be chosen as a function of the number of particles N and c_d is the volume of the unit ball in dimension d. We also define the "blob" approximation $\rho_N(t)$ to be the solution of the system (3.16) with the kernel W satisfying (3.20) given by Theorem 4.1 and "blob" initial data ρ_N^0.

In the rest, ε is chosen as a function of N as $\varepsilon(N) = N^{-\gamma/d}$ with $0 < \gamma < 1$. It is easy to check that $\|\rho_N^0\|_p \simeq N^{(\gamma-1)/p'}$ for N large enough, then we can wonder how far is the empirical measure to its blob approximation if we assume a bound on $\|\rho_N^0\|_p$ independent of N.

Proposition 5.3. *Under the assumptions of Theorem 5.1 and assuming that there exists $C_1 > 0$ independent of the number of particles N such that*

$$\|\rho_N^0\|_p \le C_1, \quad \textit{and} \quad \eta_m^0 \ge \frac{1}{C_1}\varepsilon^r,$$

with $1 \le r < \frac{d}{p'(1+\alpha)}$. Then, there exists $T > 0$ such that the solutions $\rho_N(t)$ and the empirical measure $\mu_N(t)$ are well-defined for all $t \in [0,T]$, and

$$d_\infty(\rho_N(t), \mu_N(t)) \le d_\infty(\rho_N^0, \mu_N^0) e^{C_2 T} \le \varepsilon(N) e^{C_2 T},$$

where $C_2 > 0$ is independent of N.

Proof. We follow a similar argument to Theorem 3.1. We first notice from Theorem 4.1 that there exists a common time of existence $T > 0$ of the solutions ρ_N independent of N since it only depends on $\|\rho_N^0\|_p$ and α. The empirical measure also exists up to this time since it will be smaller than the possible first collision time of particles. Moreover, due to (4.41), we get that $\|\rho_N(t)\|_p \le C$, for all $t \in [0,T]$, where C is independent of N. We next substitute $\rho_N(t)$ for $\rho(t)$ in the proof of Theorem 3.1, and thus all estimates in Step A and B hold to deduce

$$\frac{d\eta_N}{dt} \le C\eta_N \|\rho_N\| \left(1 + \eta_N^{d/p'} \eta_m^{-(1+\alpha)}\right) \le C\eta_N \left(1 + \eta_N^{d/p'} \eta_m^{-(1+\alpha)}\right),$$

and

$$\frac{d\eta_m}{dt} \ge -C\eta_m \|\rho_N\| \left(1 + \eta_N^{d/p'} \eta_m^{-(1+\alpha)}\right) \ge -C\eta_m \left(1 + \eta_N^{d/p'} \eta_m^{-(1+\alpha)}\right),$$

where $\eta_N(t) := d_\infty(\rho_N(t), \mu_N(t))$. Note that the condition $r \ge 1$ makes sense since $\varepsilon \approx \eta_N^0 \ge \eta_m^0 \ge C\varepsilon^r$ for ε small enough. We finally conclude the desired result using a similar argument as in Step C of the proof of Theorem 3.1 since

$$\frac{(\eta_N^0)^{d/p'}}{(\eta_m^0)^{1+\alpha}} \le C\varepsilon^{d/p' - r(1+\alpha)} \to 0 \quad \text{as} \quad N \to \infty,$$

by assumption. □

We now present two propositions showing that the assumptions on ρ_N^0 and η_m^0 in Proposition 5.3 are generic in a probability sense when the initial positions $X^{N,0}$ are iid with law ρ^0 in L^p. We first prove in Proposition 5.4 that η_m^0 is roughly larger than $N^{-\frac{2p-1}{d(p-1)}}$ if the $X^{N,0}$ are iid with law ρ^0.

Proposition 5.4. *Let $\rho^0 \in \mathcal{P}_1(\mathbb{R}^d) \cap L^p(\mathbb{R}^d)$, $1 < p \leq \infty$, and the initial positions $X^{N,0}$ be iid with law ρ^0. Suppose there exists $L > 0$ such that*

$$2c_d^{\frac{1}{p'}} \|\rho^0\|_p L^{\frac{d}{p'}} \leq N \,,$$

then η_m^0 satisfies

$$\mathbb{P}\left(\eta_m^0 \geq LN^{-\frac{2p-1}{d(p-1)}}\right) \geq e^{-2c_d^{\frac{1}{p'}} \|\rho^0\|_p L^{\frac{d}{p'}}} .$$

Proof. Choose an $r \in \mathbb{R}_+$. Then $\eta_m^0 \geq r$ holds if

$$X_k^0 \in \mathbb{R}^d \backslash A_k \,, \qquad \text{with} \qquad A_k = \bigcup_{1 \leq i \leq k-1} B(X_i^0, r) \,,$$

for all $1 \leq k \leq N$. It implies from our assumption with $r = LN^{-\frac{2p-1}{d(p-1)}}$ that

$$\begin{aligned} \mathbb{P}\left(\eta_m^0 \geq LN^{-\frac{2p-1}{d(p-1)}}\right) &\geq \prod_{k=1}^{N} \left[1 - \int_{A_k} \rho^0(x)\, dx\right] \\ &\geq \prod_{k=1}^{N-1} \left[1 - c_d^{\frac{1}{p'}} \|\rho^0\|_p L^{\frac{d}{p'}} N^{-2+\frac{1}{p}} k^{\frac{1}{p'}}\right], \end{aligned}$$

and thus using that $\ln(1-x) \geq -2x$ if $x \in [0, \frac{1}{2}]$, we conclude

$$\ln \mathbb{P}(\eta_m^0 \geq r) \geq -2c_d^{\frac{1}{p'}} \|\rho^0\|_p L^{\frac{d}{p'}} N^{-2+\frac{1}{p}} \sum_{k=1}^{N-1} k^{\frac{1}{p'}} \geq -2c_d^{\frac{1}{p'}} \|\rho^0\|_p L^{\frac{d}{p'}} .$$

□

The next proposition gives some bound on the large deviation of $\|\rho_N^0\|_p$. It states roughly that $\|\rho_N^0\|_p$ is of the same order that $\|\rho^0\|_p$, if the $X^{N,0}$ are iid with law ρ^0.

Proposition 5.5. *Let $\rho^0 \in \mathcal{P}_1(\mathbb{R}^d) \cap L^p(\mathbb{R}^d)$, $1 < p \leq \infty$, with compactly support included in $[-R,R]^d$. For any iid $X^{N,0}$ with law ρ^0, the smoothed empirical measures ρ_N^0 defined in (5.55) satisfy the explicit "large deviations" bound*

$$\mathbb{P}\big(L_d \|\rho^0\|_p \leq \|\rho_N^0\|_p\big) \leq [2(R+1)]^d N^\gamma e^{-c_R \|\rho\|_p N^{1-\gamma}},$$

where L_d and c_R are explicitly given by

$$c_R := \frac{2\ln 2}{[2(R+1)]^{\frac{d}{p}}} \quad \textit{and} \quad L_d := \frac{4(4[[\sqrt{d}]]+1)^{d/p}}{c_d},$$

with $[[\cdot]]$ denoting the integer part.

Proof. For any $X_i \in \mathbb{R}^d$ and $x \in \mathbb{R}^d$, we have

$$\rho_N^0(x) = \frac{1}{N\, c_d\, \varepsilon^d} \sum_{i=1}^N \mathbf{1}_{B_\varepsilon}(x - X_i) = \frac{1}{N\, c_d\, \varepsilon^d} \#\{i \text{ s.t. } |x - X_i| \leq \varepsilon\},$$

where $\#$ stands for the cardinal (of a finite set). Next, we cover $[-R, R]^d$ by M disjoint cubes C_k of size ε^d, centered at the points $(c_k)_{k\leq M}$. The number M of square needed depends on N via ε, and is bounded by

$$M \leq \left\lceil \frac{2(R+1)}{\varepsilon} \right\rceil^d.$$

Assume that $x \in C_k$ for some $1 \leq k \leq M$, i.e., $|x - c_k| \leq \frac{\sqrt{d}\varepsilon}{2}$, then

$$\#\{i \text{ s.t. } |x - X_i| \leq \varepsilon\} \leq \#\{i \text{ s.t. } |c_k - X_i|_\infty \leq 2\sqrt{d}\varepsilon\},$$

and for any $1 < p < \infty$ we obtain

$$\begin{aligned}
\int_{C_k} (\rho_N^0(x))^p\, dx &\leq \frac{\varepsilon^{d(1-p)}}{(N\, c_d)^p} \#\{i \text{ s.t. } |c_k - X_i|_\infty \leq 2\sqrt{d}\varepsilon\}^p \\
&= \frac{\varepsilon^{d(1-p)}}{(N\, c_d)^p} \#\{i \text{ s.t. } x \in C_k^d\}^p\,,
\end{aligned}$$

where C_k^d denotes the cube of center c_k and size $(4\sqrt{d}\varepsilon)^d$. Let us consider the set of cubes of the lattice that contains C_k^d, i.e.,

$$C_k^d \subset \bigcup_{j\in I_k} C_j$$

where $I_k = \{j \text{ such that } C_k^d \cap C_j \neq \emptyset\}$. It is direct to check that $\#I_k \leq M_d$ with $M_d = (4[[\sqrt{d}]]+1)^d$. Moreover, there are only M_d possible values of $1 \leq k \leq M$ such that $j \in I_k$ for a given $1 \leq j \leq M$. This yields

$$\begin{aligned}
\int_{\mathbb{R}^d} (\rho_N^0(x))^p dx &\leq \frac{\varepsilon^{d(1-p)}}{(N\, c_d)^p} \sum_{k=1}^M \sum_{j\in I_k} \#\{i \text{ s.t. } x \in C_j\}^p \\
&\leq \frac{M_d \varepsilon^{d(1-p)}}{(N\, c_d)^p} \sum_{k=1}^M \#\{i \text{ s.t. } X_i \in C_k\}^p\,. \qquad (5.56)
\end{aligned}$$

Let us introduce the notation $N_k := \#\{i \text{ s.t. } X_i \in C_k\}$. N_k is a random variable which follows a binomial law $B(N, s_k)$ with $s_k := \int_{C_k} \rho^0(x)\, dx$. If $L\|\rho^0\|_p \le \|\rho_N^0\|_p$, then (5.56) together with Hölder's inequality imply that

$$\sum_{k=1}^{M} N_k^p \ge \frac{(c_d N)^p}{M_d} \varepsilon^{d\frac{p}{p'}} \|\rho_N^0\|_p^p \ge N^p \tilde{L}^p \varepsilon^{d\frac{p}{p'}} \|\rho^0\|_p^p \ge N^p \tilde{L}^p \sum_{k=1}^{M} s_k^p,$$

where $\tilde{L} := c_d L/(M_d)^{1/p}$. But, if this happens, it means that for at least one $k \le M$,

$$\begin{aligned} N_k &\ge \left(\frac{1}{2} M^{-1} (N\tilde{L})^p \varepsilon^{d\frac{p}{p'}} \|\rho^0\|_p^p + \frac{1}{2} N^p \tilde{L}^p s_k^p \right)^{\frac{1}{p}} \\ &\ge \frac{1}{2} M^{-\frac{1}{p}} N \varepsilon^{\frac{d}{p'}} \tilde{L} \|\rho^0\|_p + \frac{1}{2} N \tilde{L} s_k \ge \frac{N\tilde{L}}{2} \left(\tilde{c}_R \varepsilon^d \|\rho^0\|_p + s_k \right), \end{aligned}$$

with $\tilde{c}_R := 1/\left[2(R+1)\right]^{\frac{d}{p}}$, where the concavity of $x^{1/p}$ was used. Then, we deduce that

$$\mathbb{P}\big(L\|\rho^0\|_p \le \|\rho_N^0\|_p\big) \le \sum_{k=1}^{M} \mathbb{P}\Big(N_k \ge \frac{N\tilde{L}}{2} [\tilde{c}_R \varepsilon^d \|\rho^0\|_p + s_k]\Big).$$

Since N_k is a random variable which follows a binomial law $B(N, s_k)$, then for any λ, the exponential moments of N_k are bounded by

$$\mathbb{E}(e^{\lambda N_k}) \le \left[1 + (e^\lambda - 1) s_k\right]^N \le e^{(e^\lambda - 1) N s_k}.$$

This together with Chebyshev's inequality implies that

$$\begin{aligned} \mathbb{P}\big(L\|\rho^0\|_p \le \|\rho_N^0\|_p\big) &\le \sum_{k=1}^{M} \mathbb{E}(e^{\lambda N_k}) e^{-\lambda \frac{N\tilde{L}}{2} [\tilde{c}_R \varepsilon^d \|\rho^0\|_p + s_k]} \\ &\le \sum_{k=1}^{M} e^{(e^\lambda - 1) N s_k - \lambda \frac{N\tilde{L}}{2} [\tilde{c}_R \varepsilon^d \|\rho^0\|_p + s_k]}. \end{aligned}$$

Taking $\lambda = \ln L'$ with the notation $L' = \frac{\tilde{L}}{2}$, we get

$$\begin{aligned} \mathbb{P}\big(L\|\rho^0\|_p \le \|\rho_N^0\|_p\big) &\le \sum_{k=1}^{M} e^{-(L' \ln L' + 1 - L') N s_k - L' \ln L' \tilde{c}_R N \varepsilon^d \|\rho^0\|_p} \\ &\le \sum_{k=1}^{M} e^{-L' \ln L' c_R N \varepsilon^d \|\rho^0\|_p} = M e^{-L' \ln L' \tilde{c}_R N \varepsilon^d \|\rho^0\|_p}, \end{aligned}$$

where we used $x \ln x - x + 1 \geq 0$, for $x > 0$. With the scaling $\varepsilon(N) = N^{-\frac{\gamma}{d}}$, we get

$$\mathbb{P}\big(L\|\rho^0\|_p \leq \|\rho_N^0\|_p\big) \leq [2(R+1)]^d N^\gamma e^{-\tilde{c}_R L' \ln L' \|\rho^0\|_p N^{1-\gamma}}.$$

In particular, choosing $L = L_d = 4(M_d)^{\frac{1}{p}}/c_d$ so that $L' = 2$, we get the desired result

$$\mathbb{P}\big(L_d\|\rho^0\|_p \leq \|\rho_N^0\|_p\big) \leq [2(R+1)]^d N^\gamma e^{-c_R\|\rho^0\|_p N^{1-\gamma}},$$

for $1 < p < \infty$. In the case of $p = \infty$, we first notice that as in (5.56), we deduce

$$\|\rho_N^0\|_\infty \leq \frac{M_d}{N c_d \varepsilon^d} \sup_{1 \leq k \leq M} \#\{i \text{ s.t. } |c_k - X_i|_\infty \leq \varepsilon\} = \frac{M_d}{N c_d \varepsilon^d} \sup_{1 \leq k \leq M} N_k.$$

Since N_k follows a binomial law $B(N, s_k)$ and $s_k \leq \|\rho^0\|_\infty \varepsilon^d$, above estimates allow us to conclude the desired inequality. □

We are now in a position to give the proof of propagation of chaos.

Proof of Theorem 5.1. We introduce several sets for the random initial data:

$$\omega_1 := \{X^{N,0} : \eta_m^0 \geq \varepsilon^r\}, \quad \omega_2 := \{X^{N,0} : L_d\|\rho^0\|_p \geq \|\rho_N^0\|_p\},$$

and

$$\omega_3 := \{X^{N,0} : d_1(\mu_N^0, \rho^0) \leq \varepsilon\},$$

where r, ε and L_d are given in Propositions 5.3, 5.4, and 5.5. We first provide the estimate of $\mathbb{P}(\omega_1^c)$. Note that since the assumption on γ, we obtain

$$\frac{2p-1}{\gamma(p-1)} < \frac{d}{p'(1+\alpha)}.$$

This yields the existence of r verifying

$$1 < \frac{2}{\gamma} \leq \frac{2p-1}{\gamma(p-1)} < r < \frac{d}{p'(1+\alpha)}.$$

This again implies the existence of $\beta > 0$ satisfying

$$\frac{d}{\gamma}\beta + \frac{2p-1}{\gamma(p-1)} < r.$$

From Proposition 5.4, if we choose $L = N^{-\beta}$, $\varepsilon = N^{-\gamma/d}$, then

$$\begin{aligned}\mathbb{P}(\omega_1^c) &= \mathbb{P}\left(X^{N,0} : \eta_m^0 \le \varepsilon^r\right) = \mathbb{P}\left(X^{N,0} : \eta_m^0 \le N^{-\frac{\gamma r}{d}}\right) \\ &\le \mathbb{P}\left(X^{N,0} : \eta_m^0 \le LN^{-\frac{2p-1}{d(p-1)}}\right) \le 1 - e^{-2c_d^{1/p'}\|\rho^0\|_p L^{d/p'}} \\ &\le 2c_d^{1/p'}\|\rho^0\|_p L^{d/p'} \le CN^{-s},\end{aligned}$$

for a sufficiently large N such that $N \ge (2c_d^{1/p'}\|\rho^0\|_p)^{\frac{p'}{p'+d\beta}}$, where $s = \frac{d\beta}{p'}$. For the estimate of $\mathbb{P}(\omega_2^c)$, we use the result of Proposition 5.5 to obtain

$$\mathbb{P}(\omega_2^c) \le CN^{\gamma}e^{-CN^{1-\gamma}}.$$

Finally the estimate of $\mathbb{P}(\omega_3^c)$ follows from (Boissard, 2011b, Proposition 1.2 of Annexe A) (see also Boissard (2011a); Bolley, Guillin, and Villani (2007)) that

$$\mathbb{P}\left(X^{N,0} : d_1(\mu_N^0, \rho^0) \ge \varepsilon\right) \le CN^{-s'},$$

where C and s' are positive constants. We now denote $\omega := \omega_1 \cap \omega_2 \cap \omega_3$. Then we have

$$\mathbb{P}(\omega^c) \le CN^{-l},$$

for some positive constants C and l. If the initial data belongs to ω, then we obtain from Proposition 5.3 that

$$d_1(\rho_N(t), \mu_N(t)) \le d_\infty(\rho_N(t), \mu_N(t)) \le \frac{Ce^{CT}}{N^{\gamma/d}}, \quad \text{for} \quad t \in [0,T].$$

We also notice from Theorem 4.1 that

$$d_1\left(\rho(t), \rho_N(t)\right) \le d_1\left(\rho^0, \rho_N^0\right)e^{CT} \le (d_1\left(\rho^0, \mu_N^0\right) + d_\infty\left(\mu_N^0, \rho_N^0\right))e^{CT},$$

for all $t \in [0,T]$. Since $d_\infty\left(\mu_N^0, \rho_N^0\right) \le \varepsilon$ and the initial data belongs to ω, this yields

$$d_1\left(\rho(t), \rho_N(t)\right) \le \frac{Ce^{CT}}{N^{\gamma/d}},$$

for all $t \in [0,T]$ since

$$d_1\left(\rho^0, \rho_{N,\varepsilon}^0\right) \le d_1\left(\rho^0, \mu_N^0\right) + d_\infty\left(\mu_N^0, \rho_{N,\varepsilon}^0\right) \le \frac{Ce^{CT}}{N^{\gamma/d}}.$$

Hence, we have

$$\mathbb{P}(\omega) \le \mathbb{P}\left(\sup_{t\in[0,T]} d_1\left(\rho(t), \rho_N(t)\right) \le \frac{Ce^{CT}}{N^{\gamma/d}}\right),$$

and it implies the desired result

$$\mathbb{P}\left(\sup_{t\in[0,T]} d_1\left(\rho(t),\rho_N(t)\right) \geq \frac{Ce^{CT}}{N^{\gamma/d}}\right) \leq \mathbb{P}(\omega^c) \leq \frac{C}{N^l}\,.$$

□

Acknowledgments

JAC was partially supported by the project MTM2011-27739-C04-02 DGI (Spain) and 2009-SGR-345 from AGAUR-Generalitat de Catalunya. JAC acknowledges support from the Royal Society by a Wolfson Research Merit Award. YPC was supported by Basic Science Research Program through the National Research Foundation of Korea funded by the Ministry of Education, Science and Technology (ref. 2012R1A6A3A03039496). JAC and YPC were supported by Engineering and Physical Sciences Research Council grants with references EP/K008404/1 (individual grant) and EP/I019111/1 (platform grant).

Bibliography

M. Agueh, R. Illner, and A. Richardson, *Analysis and simulations of a refined flocking and swarming model of Cucker-Smale type*, *Kinetic and Related Models* 4:1–16, 2011.

L.A. Ambrosio, N. Gigli, and G. Savaré, *Gradient flows in metric spaces and in the space of probability measures*, Lectures in Mathematics, Birkhäuser, 2005.

G. Albi, L. Pareschi, *Modelling self-organized systems interacting with few individuals: from microscopic to macroscopic dynamics*, *Applied Math. Letters*, 26:397–401, 2013.

I. Aoki, *A Simulation Study on the Schooling Mechanism in Fish*, Bull. Jap. Soc. Sci. Fisheries 48:1081–1088, 1982.

D. Balagué, and J. A. Carrillo, *Aggregation equation with growing at infinity attractive-repulsive potentials*, Proceedings of the 13th International Conference on Hyperbolic Problems, Series in Contemporary Applied Mathematics CAM 17, Higher Education Press, 1:136–147, 2012.

D. Balagué, Carrillo, T. J. A., Laurent, and G. Raoul, *Nonlocal interactions by repulsive-attractive potentials: radial ins/stability*, to appear in Physica D, 2013.

D. Balagué, Carrillo, T. J. A., Laurent, and G. Raoul, *Dimensionality of Local Minimizers of the Interaction Energy*, to appear in Arch. Rat. Mech. Anal., 2013.

A. Barbaro, K. Taylor, P. F. Trethewey, L. Youseff, and B. Birnir, *Discrete and continuous models of the dynamics of pelagic fish: application to the capelin*, Math. and Computers in Simulation, 79:3397–3414, 2009.

A. L. Bertozzi and T. Laurent, *Finite-time blow-up of solutions of an aggregation equation in* $\mathbb{R}^n$, Comm. Math. Phys., 274:717–735, 2007.

A. L. Bertozzi, J. A. Carrillo, and T. Laurent, *Blowup in multidimensional aggregation equations with mildly singular interaction kernels*, Nonlinearity, 22:683–710, 2009.

A. L. Bertozzi, T. Laurent, and J. Rosado, L^p *theory for the multidimensional aggregation equation*, *Comm. Pure Appl. Math.*, 43:415–430, 2010.

A. L. Bertozzi, T. Laurent, and F. Léger, *Aggregation and spreading via the newtonian potential: the dynamics of patch solutions*, Mathematical Models and Methods in Applied Sciences, 22(supp01):1140005, 2012.

M. Bodnar, J.J.L. Velazquez, *Friction dominated dynamics of interacting particles locally close to a crystallographic lattice*, accepted in Math. Methods Appl. Sci., 2012.

E. Boissard, *Simple bounds for convergence of empirical and occupation measures in 1-Wasserstein distance*, *Electron. J. Probab.* 16:2296–2333, 2011.

E. Boissard, *Problémes d'interaction discret-continu et distances de Wasserstein*, *Thesis*, http://thesesups.ups-tlse.fr/1389/, 2011.

F. Bolley, J. A. Cañizo, and J. A. Carrillo *Stochastic mean-field limit: non-Lipschitz forces & swarming*, *Math. Mod. Meth. Appl. Sci.*, 21:2179-2210, 2011.

F. Bolley, A. Guillin, and C. Villani, *Quantitative concentration inequalities for empirical measures on non-compact spaces*, *Probab. Theory Related Fields*, 137:541–593, 2007.

W. Braun and K. Hepp, *The Vlasov Dynamics and Its Fluctuations in the 1/N Limit of Interacting Classical Particles*, Commun. Math. Phys., 56:101–113, 1977.

M. Burger, P. Markowich, and J. Pietschmann, *Continuous limit of a crowd motion and herding model: Analysis and numerical simulations*, Kinetic and Related Methods, 4:1025–1047, 2011.

S. Camazine, J.-L. Deneubourg, N. R. Franks, J. Sneyd, G. Theraulaz, and E. Bonabeau, *Self-Organization in Biological Systems*, *Princeton University Press*, 2003.

J.A. Cañizo, J.A. Carrillo, and J. Rosado, *Collective Behavior of Animals: Swarming and Complex Patterns*, Arbor, 186:1035–1049, 2010.

J.A. Cañizo, J.A. Carrillo, and J. Rosado : *A well-posedness theory in measures for some kinetic models of collective motion*, Math. Mod. Meth. Appl. Sci., 21:515–539, 2011.

J. A. Carrillo, M. Di Francesco, A. Figalli, T. Laurent, and D. Slepčev, *Global-in-time weak measure solutions and finite-time aggregation for nonlocal interaction equations*, Duke Math. J., 156:229–271, 2011.

J. A. Carrillo, M. Di Francesco, A. Figalli, T. Laurent, and D. Slepčev, *Confinement in nonlocal interaction equations*, Nonlinear Anal., 75(2):550–558, 2012.

J. A. Carrillo, M. R. D'Orsogna, and V. Panferov, *Double milling in self-propelled swarms from kinetic theory*, Kinetic and Related Models 2:363-378, 2009.

J.-A. Carrillo, M. Fornasier, J. Rosado, and G. Toscani, *Asymptotic Flocking Dynamics for the kinetic Cucker-Smale model*, SIAM J. Math. Anal., 42:218–236, 2010.

J.A. Carrillo, M. Fornasier, G. Toscani, and F. Vecil, *Particle, Kinetic, and Hydrodynamic Models of Swarming*, *Mathematical Modeling of Collective Behavior in Socio-Economic and Life Sciences, Series: Modelling and Simulation in Science and Technology, Birkhauser*, pages 297-336, 2010.

J.A. Carrillo, A. Klar, S. Martin, and S. Tiwari, *Self-propelled interacting particle systems with roosting force*, Math. Mod. Meth. Appl. Sci., 20:1533–1552, 2010.

J. A. Carrillo, V. Panferov, S. Martin, *A new interaction potential for swarming models, to appear in Physica D*, 2013.

J.A. Carrillo, R.J. McCann, and C. Villani, *Kinetic equilibration rates for granular media and related equations: entropy dissipation and mass transportation estimates*, Rev. Matemática Iberoamericana, 19:1–48, 2003.

J.A. Carrillo, R.J. McCann, and C. Villani, *Contractions in the 2-Wasserstein length space and thermalization of granular media*, Arch. Rat. Mech. Anal., 179:217–263, 2006.

T. Champion, L. D. Pascale, and P. Juutinen, *The ∞-Wasserstein distance: local solutions and existence of optimal transport maps*, SIAM J. Math. Anal., 40:1–20, 2008.

I. D. Couzin, J. Krause, N. R. Franks, and S. A. Levin, *Effective leadership and decision making in animal groups on the move*, Nature 433:513–516, 2005.

F. Cucker and S. Smale, *Emergent behavior in flocks*, IEEE Trans. Automat. Control 52:852-862, 2007.

C. De Lellis and L. Székelyhidi, *The Euler equations as a differential inclusion*, Ann. of Math. 170:1417–1436, 2009.

R. Dobrushin, *Vlasov equations*, Funct. Anal. Appl. 13:115–123, 1979.

M. R. D'Orsogna, Y. L. Chuang, A. L. Bertozzi, and L. Chayes, *Self-propelled particles with soft-core interactions: patterns, stability, and collapse*, Phys. Rev. Lett. 96, 2006.

R. Eftimie, G. de Vries, and M.A. Lewis, *Complex spatial group patterns result from different animal communication mechanisms*, Proceedings of the National Academy of Sciences, 104:6974–6979, 2007.

N. Fournier, M. Hauray, and S. Mischler, *Propagation of chaos for the 2D viscous vortex model*, preprint http://arxiv.org/abs/1212.1437

I. Gallagher, L. St-Raymond, and B. Texier, *From Newton to Boltzmann: hard spheres and short-range potentials*, preprint http://arxiv.org/abs/1208.5753

C. R. Givens and R. M. Shortt, *A class of Wasserstein metrics for probability distributions*, Michigan Math. J., 31(2):231–240, 1984.

F. Golse, *The Mean-Field Limit for the Dynamics of Large Particle Systems*, Journées équations aux dérivées partielles, 9:1–47, 2003.

J. Goodman, T. Hou, and J. Lowengrub *Convergence of the point vortex method for the 2-D Euler equations*, Comm. Pure Appl. Math. 43:415–430, 1990.

C. Graham, and S. Méléard, *Stochastic particle approximations for generalized Boltzmann models and convergence estimates*, The Annals of Probability, 25:115–132, 1997.

S.-Y. Ha, J.-G. Liu, A simple proof of the Cucker-Smale flocking dynamics and mean-field limit, Commun. Math. Sci. 7 (2) (2009) 297–325.

S.-Y. Ha and E. Tadmor, *From particle to kinetic and hydrodynamic descriptions of flocking*, Kinetic and Related Models 1:415–435, 2008.

J. Haskovec, *Flocking dynamics and mean-field limit in the Cucker-Smale-type model with topological interactions*, preprint, 2013.

M. Hauray, *Wasserstein distances for vortices approximation of Euler-type equations*, Math. Mod. Meth. Appl. Sci., 19:1357–1384, 2009.

M. Hauray and P.-E. Jabin, *Particles approximations of Vlasov equations with singular forces : Propagation of chaos*, preprint.

M. Hauray and S. Mischler. *On Kac's chaos and related problems*, preprint.

C. K. Hemelrijk and H. Hildenbrandt, *Self- Organized Shape and Frontal Density of Fish Schools*, Ethology 114, 2008.

A. Huth and C. Wissel, *The Simulation of the Movement of Fish Schools*, J. Theo. Bio., 1992.

M. Kac, *Foundations of kinetic theory*, In Proceedings of the Third Berkeley Symposium on Mathematical Statistics and Probability, 1954–1955, vol. III (Berkeley and Los Angeles, 1956), University of California Press, pp. 171–197.

H. P. Mckean, *An exponential formula for solving Boltzmann's equation for a Maxwellian gas*, J. Combinatorial Theory, 2:358–382, 1967.

H. P. Mckean, *The central limit theorem for Carleman's equation*, Israel J. Math., 21:54–92, 1975.

T. Kolokonikov, H. Sun, D. Uminsky, and A. Bertozzi. *Stability of ring patterns arising from 2d particle interactions*, Physical Review E, 84:015203, 2011.

O.E. III. Lanford *Time evolution of large classical systems.* Lecture Notes in Phys. 38, Springer Verlag 1975, p. 1–111.

T. Laurent, *Local and global existence for an aggregation equation*, Communications in Partial Differential Equations, 32:1941–1964, 2007.

H. Levine, W.-J. Rappel and I. Cohen, *Self-organization in systems of self-propelled particles*, *Phys. Rev. E*, 63:017101-1/4, 2000.

Y. X. Li, R. Lukeman, and L. Edelstein-Keshet, *Minimal mechanisms for school formation in self-propelled particles*, Physica D, 237:699–720, 2008.

R. J. McCann, *Stable rotating binary stars and fluid in a tube*, Houston J. Math., 32(2):603–631, 2006.

A. Majda and A.L. Bertozzi, *Vorticity and Incompressible Flow*, Cambridge Texts in Applied Mathematics, Cambridge University Press, United Kingdom, 2002.

C. Marchioro and M. Pulvirenti, *Mathematical theory of incompressible non-viscous fluids*, Applied Mathematical Sciences 96, Springer-Verlag, New York, 1994.

S. Méléard, *Asymptotic behaviour of some interacting particle systems; McKean-Vlasov and Boltzmann models*, In Probabilistic models for non-linear partial differential equations (Montecatini Terme, 1995), Lecture Notes in Math. 1627. Springer, Berlin, 1996.

S. Mischler and C. Mouhot, *Kac's program in kinetic theory*, to appear in Inventiones mathematicae, 2013.

A. Mogilner and L. Edelstein-Keshet, *A non-local model for a swarm*, *J.* Math. Bio., 38:534–570, 1999.

A. Mogilner, L. Edelstein-Keshet, L. Bent, and A. Spiros, *Mutual interactions, potentials, and individual distance in a social aggregation*, *J.* Math. Biol., 47:353–389, 2003.

H. Neunzert, *An introduction to the nonlinear Boltzmann-Vlasov equation*, In Kinetic theories and the Boltzmann equation (Montecatini Terme, 1981), Lecture Notes in Math. 1048. Springer, Berlin, 1984.

H. Osada, *Propagation of chaos for the two-dimensional Navier-Stokes equation*, In Probabilistic methods in mathematical physics (Katata/Kyoto, 1985), Academic Press, Boston, 1987, 303–334.

F. Otto, The geometry of dissipative evolution equations: the porous medium equation, Comm. Partial Differential Equations 26:101–174, 2001.

J. Parrish, and L. Edelstein-Keshet, *Complexity, pattern, and evolutionary trade-offs in animal aggregation*, Science, 294: 99–101, 1999.
L. Perea, G. Gómez, P. Elosegui, *Extension of the cucker–smale control law to space flight formations*, AIAA Journal of Guidance, Control, and Dynamics, 32:527–537, 2009.
S. Schochet, *The point-vortex method for periodic weak solutions of the 2-D Euler equations*, Comm. Pure Appl. Math. 49:911–965, 1996.
H. Spohn, *Large scale dynamics of interacting particles*, Texts and Monographs in Physics, Springer, 1991.
A.-S. Sznitman, *Topics in propagation of chaos*, In Ecole d'Eté de Probabilités de Saint-Flour XIX 1989, Lecture Notes in Math. 1464. Springer, Berlin, 1991.
C.M. Topaz and A.L. Bertozzi, *Swarming patterns in a two-dimensional kinematic model for biological groups*, SIAM J. Appl. Math., 65:152–174, 2004.
C.M. Topaz, A.L. Bertozzi, and M.A. Lewis, *A nonlocal continuum model for biological aggregation*, Bulletin of Mathematical Biology, 68:1601–1623, 2006.
T. Vicsek, A. Czirok, E. Ben-Jacob, I. Cohen, and O. Shochet, *Novel type of phase transition in a system of self-driven particles*, Phys. Rev. Lett., 75:1226–1229, 1995.
C. Villani, *Limite de champ moyen*, Lecture notes, 2002.
C. Villani, *Topics in optimal transportation*, volume 58 of Graduate Studies in Mathematics, American Mathematical Society, Providence, RI, 2003.

Non smooth evolution models in crowd dynamics: mathematical and numerical issues

Bertrand Maury*

* Laboratoire de Mathématique d'Orsay, Université Paris-Sud, 91405 Orsay, France

Abstract Accounting for hard congestion in crowd motion modeling leads to non-smooth evolution problems. At the microscopic level (individuals are represented separately), these problems fit in the framework of non smooth analysis in Hilbert spaces, and the tools developed in the 70's to handle the so-called *sweeping* process are directly adaptable. At the macroscopic scale (the population is represented by a density), a similar approach can be carried out. This is done by identifying densities with measures in the Wasserstein space, endowed with the distance based on optimal transportation for the quadratic cost. These lecture notes provide an introduction to the mathematical theory of these models, and a description of numerical methods, in the microscopic (ODE) and macroscopic (PDE) cases.

1 Introduction

Congestion is a crucial issue in crowd motion modeling. When people want to evacuate a room they all head to the door, which tends to decrease their mutual distance, thereby increasing the local density. Investigating a bit further these considerations will shed light on the different aspects of congestion, in the microscopic and in the macroscopic settings. As a first step, consider people in a corridor, and assume that they all want to go to the right direction. The situation is likely to become critical in terms of density if the velocities decrease from left to right (i.e. the people in front move slower than the people behind them), whereas increasing velocities will tend to relax the situation (the density decreases). In the two-dimensional setting, the notion of "increasing" velocity field has to be extended in some way. Consider a velocity field $\mathbf{U}$ in the plane, and two individuals 1 and 2, identified to rigid discs centered at $\mathbf{q}_1$ and $\mathbf{q}_2$. Each individual $\mathbf{q}_i$ is assumed to move at velocity $\mathbf{U}(\mathbf{q}_i)$. Their distance increases whenever

$$(\mathbf{U}(\mathbf{q}_2) - \mathbf{U}(\mathbf{q}_1)) \cdot (\mathbf{q}_2 - \mathbf{q}_1) \geq 0. \tag{1}$$

A velocity field that verifies (1) is said to be *monotone* (which actually means increasing in a general, multidimensional sense). An archetypal monotone field is $\mathbf{x}/|\mathbf{x}|$. The velocity field corresponding to the evacuation of a room (the desired velocity is directed to a single "point", which is the door) is roughly the opposite of this field: the evolution according to this field tends to decrease all distances. Let us go a bit further in the classification between relaxing and non relaxing fields. The monotony (1) condition above can be written for $\mathbf{q}_2$ seen as a variation of $\mathbf{q}_1$, i.e. $\mathbf{q}_2 = \mathbf{q}_1 + \varepsilon\mathbf{h}$. Having ε go to 0, we obtain that

$$\mathbf{h} \cdot \nabla \mathbf{U} \cdot \mathbf{h} \geq 0$$

for any direction $\mathbf{h}$. Note that it can be expressed in terms of the symmetrized gradient of $\mathbf{U}$, i.e. $\nabla \mathbf{U} + {}^t\nabla \mathbf{U}$, since the skew symmetric part does not contribute. Now consider particles (or individuals) evolving according to this velocity fields. The velocity of a point $\mathbf{q} + \varepsilon\mathbf{h}$ close to a reference point $\mathbf{q}$ writes

$$\mathbf{U}(\mathbf{q} + \varepsilon\mathbf{h}) = \mathbf{U}(\mathbf{q}) + \varepsilon \left(\frac{\nabla \mathbf{U} - {}^t\nabla \mathbf{U}}{2} \right) \cdot \mathbf{h} + \varepsilon \left(\frac{\nabla \mathbf{U} + {}^t\nabla \mathbf{U}}{2} \right) \cdot \mathbf{h} + o(\varepsilon),$$

which means that the local velocity around $\mathbf{q}$ is mainly a rigid motion composed of a translation (velocity $\mathbf{U}(\mathbf{q})$), a rotation (angular velocity $\partial_1 u_2 - \partial_2 u_1$), and a last component which corresponds to deformations. The corresponding tensor (or matrix)

$$\mathbf{e} = \frac{\nabla \mathbf{U} + {}^t\nabla \mathbf{U}}{2} \tag{2}$$

is called the *strain tensor*. It is symmetric, thereby diagonalizable, with real eigenvalues, the sign of which indicate the nature of the deformation: a negative value correspond to compression in the corresponding direction, whereas a positive one reflects expansion (i.e. distances increase). Condition (1) therefore expresses that all eigenvalues of $\mathbf{e}$ are nonnegative, which means that the flow is expanding *in all directions*. Coming back to the problem of congestion, it means that problems can be expected (i.e. some distances between individuals may decrease down to physical contact) as soon as the desired velocity field is not expanding in all directions. Actually, the "problems" we mentioned concern safety (i.e. jamming can be expected upstream the exit, which may tend to increase the evacuation time and induce casualties), but the fact that the field concentrates people makes the problem *easier* from the mathematical standpoint, as far as the uncontested situation is concerned. Indeed, if $-\mathbf{U}$ is monotone, as it is expected for an evacuation through a small exit, then the evacuation process (without congestion) can be written

$$\frac{d\mathbf{q}}{dt} - \mathbf{U} \ni 0,$$

and well-posedness (existence and uniqueness of a solution) can be established without assuming that $\mathbf{U}$ is Lipschitz, as the standard theory of Ordinary Differential Equations requires (see e.g. Bauschke and Combettes (2011)).

Let us now point out a deep difference between the microscopic approach that we followed previously and the macroscopic one. The latter consists in representing the population by a density $\rho(x,t)$. Still denoting by $\mathbf{U}$ the underlying velocity field[1], the "people conservation" in any domain ω expresses the time derivative of the population in ω and the flux through its boundary, and it writes

$$\frac{d}{dt}\int_\omega \rho = \int_\omega \frac{\partial \rho}{\partial t} = -\int_{\partial\omega} \rho\mathbf{U}\cdot\mathbf{n} = -\int_\omega \nabla\cdot(\rho\mathbf{U}).$$

Since conservation holds for any such subdomain, we have the transport equation

$$\frac{\partial \rho}{\partial t} + \nabla\cdot(\rho\mathbf{U}) = 0.$$

The derivative of ρ along a trajectory (total derivative) writes

$$\frac{\partial \rho}{\partial t} + \mathbf{U}\cdot\nabla\rho = -\rho\nabla\cdot\mathbf{U},$$

which is nonpositive (i.e. the density relaxes toward a smaller value) as soon as $\nabla\cdot\mathbf{U} \geq 0$. Note that the condition is weaker than the microscopic one: preserving nondecreasing distance necessitates expansion in all directions (the eigenvalues of $\mathbf{e}$ are nonnegative), whereas here only the sign of the trace (that is the divergence of $\mathbf{U}$, and also the sum of eigenvalues of the strain tensor $\mathbf{e}$ defined by (2)) is relevant. As we shall see in Section 2, handling the constraints at the microscopic level consists in accounting for the concentrating character of the velocity field, neither in all directions (as in the first approach we presented), nor in a mean sense like in the macroscopic setting, but rather *in some particular directions* that correspond to the actual contacts between individuals. Those directions depend

[1] We shall make later on a difference between the *desired* velocity field and the *actual* velocity field; the latter accounts for constraints.

on the local structure of the contact network, and the native heterogeneity of jammed population will induce numerical difficulties. The macroscopic problem is easier from this standpoint, since the problem is homogeneous: the constraint to remain below a prescribed value is the same over the whole saturated zone. The difficulty here is rather due to the Eulerian character of Partial Differential Equations, which rules out the possibility to directly apply tools of convex analysis, at least in standard functional spaces. We shall details in Section 3 how the *Wasserstein Distance* on measures will make it possible to adapt some tools that have been developped in Hilbert spaces.

2 Hard congestion in the microscopic setting

2.1 The model

We consider N individuals identified to rigid disks of radius r in a room identified to a domain Ω. The positions are represented by a vector $\mathbf{q}$ of $\mathbb{R}^{2N}$:

$$\mathbf{q} = (\mathbf{q}_1, \mathbf{q}_2, \cdots, \mathbf{q}_N) \in \mathbb{R}^{2N}.$$

The distance between two individuals i and j is denoted by

$$D_{ij} = |\mathbf{q}_j - \mathbf{q}_i| - 2r.$$

The set of feasible configurations (contact is allowed, but no overlapping) can be written

$$K = \left\{\mathbf{q} \in \mathbb{R}^{2N}, \ D_{ij} \geq 0 \quad \forall i \neq j\right\}.$$

The distance D_{ij} can be considered as a function of the configuration vector $\mathbf{q}$ (although it depends on i and j only), and we denote by $\mathbf{G}_{ij} \in \mathbb{R}^{2N}$ its gradient. Ruling out overlapping between two individual amounts to require that their instantaneous relative velocity is nonnegative in the normal direction, as soon as there is contact. It leads to the following definition of the set of feasible (or admissible) velocities[2]

$$C_K(\mathbf{q}) = \left\{\mathbf{v} \in \mathbb{R}^{2N}, \ D_{ij}(\mathbf{q}) = 0 \Longrightarrow \mathbf{G}_{ij} \cdot \mathbf{v} \geq 0\right\}. \tag{3}$$

Now consider that a collection of desired velocities

$$\mathbf{U} = (\mathbf{U}_1, \ldots, \mathbf{U}_N)$$

[2] In order to alleviate notations, we disregard here the constraints that are due to obstacles (walls or piles), but they can be integrated in the set of feasible velocities like inter individual non overlapping constraints.

is given. Following Maury and Venel (2011), we assume that, at each instant, the actual velocity field $\mathbf{u}$ is the closest (in the ℓ^2 sense) to the desired velocity field $\mathbf{U}$, among all feasible fields, i.e.

$$\mathbf{u} = \frac{d\mathbf{q}}{dt} = P_{C_K(\mathbf{q})}(\mathbf{U}). \tag{4}$$

The desired velocity $\mathbf{U}_i$ may depend on time, on the position of other individuals (if one aims at accounting for social effects). If one considers that the desired velocity of an individual depends on its location only, and that individual are interchangeable, $\mathbf{U}_i$ can be defined as $\mathbf{U}_0(\mathbf{q}_i)$, where $\mathbf{U}_0$ is a desired velocity field, the same for everybody. We consider here the general case: the individual desired velocity depends on all the positions, and this dependence may vary from an individual to the other. Thus, the desired velocity is written $\mathbf{U} = \mathbf{U}(\mathbf{q})$.

2.2 General setting, catching up approach

The basic tool to reformulate Model (4) is the decomposition of a Hilbert space according to mutually polar cones, as proposed by Moreau (1962). Consider a closed convex cone C pointed at the origin, in a Hilbert space H, i.e.

$$C \subset H\,,\ \mathbb{R}^+C \subset C\,,\ \lambda x + (1-\lambda)y \in C \quad \forall x, y \in C\,,\ \lambda \in [0,1]\,,\ \overline{C} = C.$$

The polar cone to C is defined as

$$C^\circ = \{y \in H\,,\ (y,x) \le 0 \quad \forall x \in C\}\,.$$

The decomposition of any Hilbert space as the direct sum of a closed vector set and its orthogonal can be extended to cones (Moreau (1962)):

$$I_d = P_C + P_{C^\circ},$$

where I_d is the identity, and P_C (resp. P_{C°) is the projection on C (resp. C°).

Eq. (4) can be rewritten

$$\mathbf{u} = \frac{d\mathbf{q}}{dt} = \mathbf{U} - P_{C_K^\circ(\mathbf{q})}(\mathbf{U}), \tag{5}$$

which implies $\mathbf{u} - \mathbf{U} \in -C_K^\circ(\mathbf{q})$. The latter inclusion is usually written

$$\frac{d\mathbf{q}}{dt} + N_K(\mathbf{q}) \ni \mathbf{U}(\mathbf{q}), \tag{6}$$

where $N_K = C_K^\circ(\mathbf{q})$ is the so-called *outward normal cone* to $\mathcal{K}$ at $\mathbf{q}$. As we shall see, this inclusion actually characterizes the evolution (i.e. no information is lost by replacing the projection by a simple inclusion).

Catching up algorithm in the convex case Now assume for a moment that the set K of feasible configurations is convex[3]. By characterization of the projection on a closed convex set, i.e.

$$\mathbf{q} = P_K\tilde{\mathbf{q}} \iff (\tilde{\mathbf{q}} - \mathbf{q}, \mathbf{z} - \mathbf{q}) \leq 0 \quad \forall \mathbf{z} \in K,$$

we obtain that the outward normal cone $N_K(\mathbf{q}) = C_K^\circ(\mathbf{q})$ can be expressed

$$N_K(\mathbf{q}) = \{\tilde{\mathbf{q}} - \mathbf{q}\,,\ \mathbf{q} = P_K\tilde{\mathbf{q}}\}\,. \tag{7}$$

This suggests the following time discretization, which can be seen as a semi-implicit Euler scheme applied to (5). This strategy was introduced by Moreau (1977) as a *catching-up* approach to build solutions to similar problems (sweeping process).

Considering a time step $\tau > 0$, it consists in discretizing (5) as

$$\frac{\mathbf{q}^{n+1} - \mathbf{q}^n}{\tau} - \mathbf{U}(\mathbf{q}^n) \in -N_K(\mathbf{q}^{n+1}), \tag{8}$$

which can be written

$$\mathbf{q}^n + \tau\mathbf{U}(\mathbf{q}^n) - \mathbf{q}^{n+1} \in N_K(\mathbf{q}^{n+1}).$$

By (7), it is equivalent to (Catching-up scheme)

$$\mathbf{q}^{n+1} = P_K\left(\mathbf{q}^n + \tau\mathbf{U}(\mathbf{q}^n)\right), \tag{9}$$

so that the semi-*implicit* Euler scheme turns out to be *explicit* in this regard: it reduces to a projection of a predicted position $\tilde{\mathbf{q}}^{n+1} = \mathbf{q}^n + \tau\mathbf{U}(\mathbf{q}^n)$ on K.

Theoretical issues (convex case) The catching-up algorithm can be used to build a solution to our problem, more precisely its formulation (6). We shall state it in an abstract form.

Proposition 2.1. *Let H be a Hilbert space, $K \subset H$ a closed convex set, and*

$$\mathbf{U}\ :\ \mathbf{q} \in H \longmapsto \mathbf{U}(\mathbf{q})$$

[3]This assumption is very strong, and rarely verified in realistic situations in the context of crowd motions. It corresponds to the case of a single person in a convex room with no door (!), or to the one-dimensional situation: N individual in a corridor, the size of which is exactly the diameter of individuals.

a Lipschitz function in H*, bounded over* H*. Consider an initial condition* $\mathbf{q}^0$*,* $T > 0$ $\tau = T/M > 0$ *a time step, and*

$$\mathbf{q}_\tau^0\,,\ \mathbf{q}_\tau^1\,,\ \ldots\,,\ \mathbf{q}_\tau^M\,,$$

the elements obtained by application of the Catching-Up algorithm (9)*. Denoting by* $\mathbf{q}_\tau$ *the corresponding continuous, piecewise affine trajectory,* $\mathbf{q}_\tau$ *converges uniformly toward a solution* $t \mapsto \mathbf{q}(t)$ *to* (6)*, i.e.*

$$\frac{d\mathbf{q}}{dt} + N_K(\mathbf{q}) \ni \mathbf{U}(\mathbf{q}(t)), \textit{for almost every } t \in (0,T).$$

Proof. We shall simply give here the main arguments of the proof, and we refer to Venel (2011) for further details. The main argument is a characterization of the outward normal cone for a convex set K (see e.g. Venel (2011)):

$$\begin{aligned} \mathbf{v} \in N_K(\mathbf{q}) \quad &\Longleftrightarrow \quad \forall \boldsymbol{\xi}\,,\ \langle \mathbf{v}\,,\boldsymbol{\xi}\rangle \le |\mathbf{v}|\, d_K(\mathbf{q}+\boldsymbol{\xi}) \\ &\Longleftrightarrow \quad \exists C>0\,,\ \eta>0\,,\ \forall \boldsymbol{\xi}\,,\ |\boldsymbol{\xi}| \le \eta\,,\ \langle \mathbf{v}\,,\boldsymbol{\xi}\rangle \le C d_K(\mathbf{q}+\boldsymbol{\xi}), \end{aligned} \tag{10}$$

where $d_K(\cdot)$ denotes the distance to K.

The rest of the proof relies on compactness arguments. First of all $(\mathbf{q}_\tau)$ is uniformly bounded in $W^{1,\infty}(0,T)$, so that (up to the extraction of a subsequence) $(\mathbf{q}_\tau)_\tau$ converges uniformly toward some trajectory $\mathbf{q}$, and the corresponding velocity $\mathbf{u}_\tau = d\mathbf{q}_\tau/dt$ (which is piecewise constant) converges toward some $\mathbf{u}$ in the L^∞ weak-star topology.

It holds that

$$\mathbf{q}_\tau^{n+1} = P_K\left(\mathbf{q}_\tau^n + \tau \mathbf{U}(\mathbf{q}_\tau^n)\right),$$

which implies

$$(\mathbf{q}_\tau^n + \tau \mathbf{U}(\mathbf{q}_\tau^n)) - \mathbf{q}_\tau^{n+1} \in N_K(\mathbf{q}_\tau^{n+1}),$$

i.e.

$$\frac{d\mathbf{q}_\tau}{dt} - \mathbf{U}(\tilde{\mathbf{q}}_\tau) \in -N_K(\overline{\mathbf{q}}_\tau), \text{ for a.e. } t \in (0,T),$$

where $\tilde{\mathbf{q}}_\tau$ and $\mathbf{q}_\tau$ are piecewise constant approximate solutions, with value at the beginning and at the end of time intervals, respectively. Notice that both sequences $\tilde{\mathbf{q}}_\tau$ and $\overline{\mathbf{q}}_\tau$ uniformly converge to $\mathbf{q}$. By (10), the previous inclusion implies

$$-\left\langle \frac{d\mathbf{q}_\tau}{dt} - \mathbf{U}(\tilde{\mathbf{q}}_\tau)\,,\boldsymbol{\xi}\right\rangle \le \underbrace{\left|\frac{d\mathbf{q}_\tau}{dt} - \mathbf{U}(\tilde{\mathbf{q}}_\tau)\right|}_{\le M} d_K\left(\overline{\mathbf{q}} + \boldsymbol{\xi}\right).$$

Since $\mathbf{u}_\tau - \mathbf{U}(\tilde{\mathbf{q}}_\tau)$ converges weakly-star to $\mathbf{u} - \mathbf{U}(\mathbf{q}_\tau)$ (in L^∞), by Mazur's Lemma, a convex combination $\mathbf{z}_\tau$ of the sequence $(\mathbf{u}_\tau - \mathbf{U}(\tilde{\mathbf{q}}_\tau))$ converges strongly in L^1 toward $\mathbf{u} - \mathbf{U}(\mathbf{q})$. We therefore have point wise convergence of $\mathbf{z}_\tau$ toward $\mathbf{u} - \mathbf{U}(\mathbf{q})$ for almost every time t. Thus, for any time at which pointwise convergence of $\mathbf{z}_\tau$ holds,

$$\limsup \langle \mathbf{z}_\tau \, , \boldsymbol{\xi} \rangle \leq M d_K \left(\mathbf{q} + \boldsymbol{\xi} \right).$$

so that.

$$\left\langle \frac{d\mathbf{q}}{dt} - \mathbf{U}(\mathbf{q}) \, , \boldsymbol{\xi} \right\rangle \leq M d_K \left(\mathbf{q} + \boldsymbol{\xi} \right).$$

By using again (10) (second characterization) the other way around, we obtain

$$\frac{d\mathbf{q}}{dt} - \mathbf{U}(\mathbf{q}) \in -N_K(\mathbf{q}) \quad \text{for a.e. } t,$$

which ends the proof. □

Crowd motion model: the non convex situation As for the crowd motion model we introduced, the feasible set K is not convex as soon as there are two individuals (i.e. rigid discs). Yet, as suggested by the previous approach, similar results can be expected as soon as it is possible to project the predicted configuration $\tilde{\mathbf{q}}^{n+1} = \mathbf{q}^n + \tau \mathbf{U}(\mathbf{q}^n)$ on K. More precisely, if $\mathbf{q}^n$ is feasible (i.e. in K), if $\mathbf{U}$ is bounded, then a control on the time step ensures that $\tilde{\mathbf{q}}^{n+1}$ is close to K. It is therefore enough to prescribe that the projection is well-defined in the neighborhood of K. This leads to the notion of *prox-regular* sets (Poliquin and Rockafellar (1996)). Let us first extend the notion of outward normal cone to non convex sets (Clarke et al. (1995)):

Definition 2.2. Let $K \subset H$ be a closed set, and $\mathbf{q} \in K$. The outward normal cone to K at $\mathbf{q}$ is defined by

$$N_K(\mathbf{q}) = \{ \mathbf{v} \in H \, , \; \exists \alpha > 0 \, , \; \mathbf{q} \in P_K(\mathbf{q} + \alpha \mathbf{v}) \} \, , \tag{11}$$

where $P_K(\tilde{\mathbf{q}})$ denotes here the set of elements of H that realize the distance between $\tilde{\mathbf{q}}$ and K.

The prox-regularity is defined as follows:

Definition 2.3. Let $K \subset H$ be a closed set, and $\eta > 0$. The set K is said to be uniformly η prox-regular if, for any $\mathbf{q} \in \partial K$, $N_K(\mathbf{q})$ is not reduced to $\{0\}$ and, for any $\mathbf{v} \in N_K(\mathbf{q})$, with $|\mathbf{v}| = 1$, we have

$$B(\mathbf{q} + \eta \mathbf{v}, \eta) \cap K = \emptyset.$$

A weakened form of the characterization (10) holds for prox-regular sets:

$$\mathbf{v} \in N_K(\mathbf{q}) \Longleftrightarrow$$

$$\exists \alpha > 0\,,\; C > 0\,,\; \forall \boldsymbol{\xi}\,,\; |\boldsymbol{\xi}| < \alpha\,,\; \langle \mathbf{v}\,, \boldsymbol{\xi} \rangle \le C d_K(\mathbf{q} + \boldsymbol{\xi}) + \frac{|\mathbf{v}|}{2\eta} |\boldsymbol{\xi}|^2\,, \qquad (12)$$

Thanks to this characterization, Prop. 2.1 can be extended to the prox regular case, which provides a framework for the crowd motion model, since the feasible set K can be shown to be prox-regular (Maury and Venel (2011)).

Saddle point formulation The outward normal cone $N_K(\mathbf{q}) = C_K^\circ(\mathbf{q})$ can be parametrized thanks to Farkas' Lemma (see e.g. Rockafellar (1970), p. 200):

$$\begin{aligned} N_K(\mathbf{q}) &= \left\{ \mathbf{v} \in \mathbb{R}^{2N}\,,\; D_{ij}(\mathbf{q}) = 0 \Longrightarrow \mathbf{G}_{ij} \cdot \mathbf{v} \ge 0 \right\}^\circ \\ &= \left\{ -\sum_{i<j} \lambda_{ij} \mathbf{G}_{ij}\,,\; \lambda_{ij} \ge 0\,,\; D_{ij} > 0 \Rightarrow \lambda_{ij} = 0 \right\}. \end{aligned}$$

As a consequence, the projection of $\mathbf{U}$ on $N_K(\mathbf{q})$ can be formulated in a saddle point manner. It amounts to find $(\lambda_{ij})_{(i,j)\in\Lambda}$, where Λ is the set of (i,j) such that the constraint is active (i.e. $D_{ij}(\mathbf{q}) = 0$), and $\mathbf{u} \in \mathbb{R}^{2N}$, such that

$$\left\{ \begin{array}{lcl} \mathbf{u} + B^\star \lambda & = & \mathbf{U} \\ B\mathbf{u} & \le & 0 \\ \lambda & \ge & 0 \\ (B\mathbf{u}, \lambda) & = & 0. \end{array} \right. \qquad (13)$$

where B is a matrix, each line of which corresponds to the constraint

$$-\mathbf{G}_{ij} \cdot \mathbf{u} \le 0, \quad (i,j) \in \Lambda,$$

with

$$\mathbf{G}_{ij} = (0, \ldots, 0, -\mathbf{e}_{ij}, 0, \ldots, 0, \mathbf{e}_{ij}, 0, \ldots, 0)\;,\; \mathbf{e}_{ij} = \frac{\mathbf{q}_j - \mathbf{q}_i}{|\mathbf{q}_j - \mathbf{q}_i|}.$$

Thus, the projection takes the form of a (unilateral) discrete Darcy problem. This analogy is not only formal. Consider for example the case of people in a corridor (one-dimensional setting):

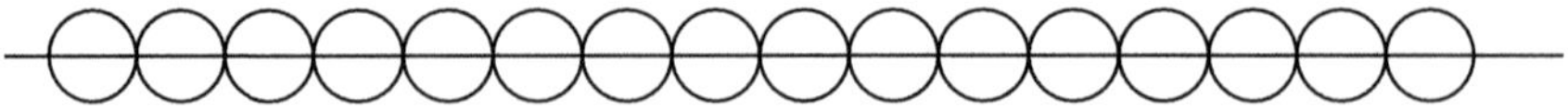

The gradients are

$$\mathbf{G}_{12} = (-1, 1, 0, \ldots, 0) \ , \ \mathbf{G}_{23} = (0, -1, 1, 0, \ldots, 0) \ , \quad \text{etc.}$$

so that the constraint matrix writes

$$B = \begin{pmatrix} 1 & -1 & 0 & \ldots & \ldots \\ 0 & 1 & -1 & \ldots & \ldots \\ 0 & 0 & \ddots & \ddots & \ldots \\ 0 & 0 & \ldots & 1 & -1 \end{pmatrix}$$

that is the discrete counterpart of $-\partial_x$ (opposite of the divergence), and $B^\star$ corresponds to ∂_x (gradient). Note that $BB^\star$ is the discrete counterpart of the one-dimensional Laplacian $-\partial_{xx}$. This observation reveals the numerical underlying difficulties: at each time step, the computation of the actual velocity can be expected to be, at least, as difficult as solving a discrete Poisson problem with the same number of degrees of freedom. It also illustrates the non-local effect of the projection onto the set of feasible velocities: all individuals gathered in a same cluster are likely to interact.

The projection (9) can also be formulated in a dual manner: since

$$\mathbf{q}^{n+1} = P_K \left(\mathbf{q}^n + \tau \mathbf{U}(\mathbf{q}^n) \right)$$

there exists a collection of Lagrange multipliers $(\lambda_{ij})_{(i,j) \in \Lambda}$ such that

$$\mathbf{q}^n + \tau \mathbf{U}(\mathbf{q}^n) = \mathbf{q}^{n+1} - \sum_{i,j} \lambda_{ij} \mathbf{G}_{ij}(\mathbf{q}^{n+1}),$$

which can be written

$$\frac{\mathbf{q}^{n+1} - \mathbf{q}^n}{\tau} + B^\star \lambda = \mathbf{U}(\mathbf{q}^n).$$

Yet, in spite of its formal simplicity, this formulation does not directly lead to a tractable numerical scheme, since the matrix B (together with the set Λ of active couples) depends on the unknown $\mathbf{q}^{n+1}$. It reflects the implicit (and highly nonlinear) character of the scheme.

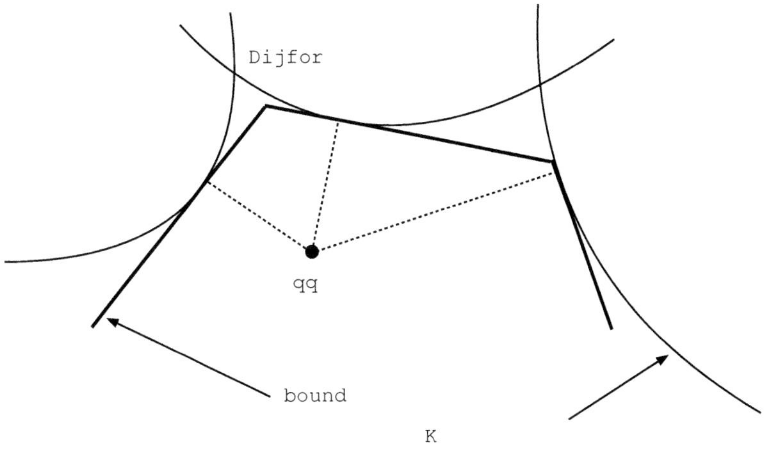

Figure 1. Inner approximation of the feasible set K.

2.3 Numerical scheme

In order to obtain a tractable numerical scheme, we replace (as in Maury (2006)) the set of feasible configurations by some kind of local inner approximation. More precisely, considering a given configuration $\mathbf{q} \in K$, we introduce (see Fig. 1)

$$\tilde{K}_{\mathbf{q}} = \{\tilde{\mathbf{q}}\,,\ D_{ij}(\mathbf{q}) + \mathbf{G}_{ij}(\mathbf{q}) \cdot (\tilde{\mathbf{q}} - \mathbf{q}) \geq 0 \quad \forall i < j\}\,.$$

Thanks to the convex character of the function D_{ij}, it can be shown that $\tilde{K}_{\mathbf{q}} \subset K$ for any $\mathbf{q} \in K$, and it is a convex polyhedron as intersection of half spaces. The projection step (9) is replaced by

$$\mathbf{q}^{n+1} = P_{\tilde{K}_{\mathbf{q}^n}} \left(\mathbf{q}^n + \tau \mathbf{U}(\mathbf{q}^n)\right), \tag{14}$$

which amounts to project a predicted position $\tilde{\mathbf{q}}^{n+1} = \mathbf{q}^n + \tau\mathbf{U}(\mathbf{q}^n)$ on the set $\tilde{K}_(\mathbf{q}^n)$, which is an intersection of half spaces. The link with the initial problem is more explicit if one expresses the scheme in terms of velocities. The approximate set of feasible velocities is defined by

$$\tilde{C}^\tau(\mathbf{q}) = \{\mathbf{v}\,,\ D_{ij}(\mathbf{q}) + \tau \mathbf{G}_{ij}(\mathbf{q}) \cdot \mathbf{v} \geq 0 \quad \forall i < j\}\,.$$

Now setting $\mathbf{u}^{n+1} = (\mathbf{q}^{n+1} - \mathbf{q}^n)/\tau$, the scheme (14) can be expressed in terms of velocities as

$$\mathbf{u}^{n+1} = P_{\tilde{C}^\tau(\mathbf{q}^n)} \mathbf{U}(\mathbf{q}^n), \tag{15}$$

which the discrete counterpart to (4). The problem can be put in a saddle-point form:

$$\left\{\begin{array}{lcl} \mathbf{u} + B^\star\lambda & = & \mathbf{U} \\ B\mathbf{u} & \leq & D/\tau \\ \lambda & \geq & 0 \\ (B\mathbf{u} - D/\tau, \lambda) & = & 0. \end{array}\right. \tag{16}$$

where D is the vector of distances at the current configuration, i.e. $(D_{ij}(\mathbf{q}^n))$.

Remark 2.4. In the present context, the number of rows of B is the number of *potential* contacts, i.e. $N(N-1)/2$), whereas the number of rows of B in the continuous setting (13) was the number of *actual* contacts, which is of the order of $3N$. It is nevertheless possible to alleviate the computational costs by ruling out *a priori* the constraints that are not likely to become activated at the next step, which amounts to suppress the corresponding lines of B, thereby reducing the computational costs.

As it appears in Fig. 1, K can hardly be seen as a global approximation of $\tilde{K}_{\mathbf{q}}$. Yet, if the time step is small, the constraints will be activated when $\mathbf{q}$ is closed to a forbidden zone, and in this case $\tilde{K}_{\mathbf{q}}$ approximates K locally, i.e. in the neighborhood of $\mathbf{q}$. It can be proven that the scheme converges (Venel, 2011).

Uzawa algorithm Solutions to Problem (16) can be approximated by the Uzawa algorithm. Let λ^0 be given (the choice $\lambda = 0$ can be made in case no prior information on λ is available). Successive approximations $(\mathbf{u}^k, \lambda^k)$ are built as follows: once λ^k is determined, $\mathbf{u}^{k+1}$ and λ^{k+1} are defined by

$$\left\{\begin{array}{lcl} \mathbf{u}^{k+1} + B^\star\lambda^k & = & \mathbf{U} \\ \lambda^{k+1} & = & \Pi_+\left(\lambda^k + \nu\left(B\mathbf{u}^{k+1} - D/\tau\right)\right) \end{array}\right. \tag{17}$$

where Π_+ is the euclidean projection on the cone of vectors with nonnegative components (a simple cut-off in practice), and $\nu > 0$ is a fixed parameter.

Note that any fixed point of the algorithm is a solution to Problem (16). Indeed, consider $(\mathbf{u}, \lambda)$ such a fixed point,. For any component of λ associated to a contact (i, j), stationarity implies the following alternative:

i) either $\lambda_{ij} = 0$ and $(B\mathbf{u} - D/\tau)_{ij} \leq 0$ (the constraint is satisfied, possibly in a strict sense, and the Lagrange multiplier is inactive), or

ii) $\lambda_{ij} > 0$ and $(B\mathbf{u} - D/\tau)_{ij} = 0$ (the constraint is saturated, and the Lagrange multiplier is active).

The algorithm can be shown to converge as soon as $0 < \nu < 2/\|B\|^2$ (see Ciarlet (1989)). This algorithm is actually a fixed-step algorithm of the projected gradient type, performed on the quadratic functional defined on the set of Lagrange multipliers:

$$\mu \longmapsto \Psi(\mu) = \frac{1}{2}\left(\mathbf{U} - B^\star\mu, \mathbf{U} - B^\star\mu\right) + \left(\mu, D/\tau\right).$$

The quadratic part of the functional is the quadratic form associated to the matrix $BB^\star$, that is the discrete Laplace-like operator on the network already mentioned in the previous section. This very matrix will condition the numerical difficulty so numerically solve the system. Note that the present matrix $BB^\star$ actually differs from the one described in the previous section in that it pertains to all potential contacts, and not only actual ones (see Remark 2.4). Yet, in actual computations, active contacts for the discretization scheme will correspond to couples of particles that are not far away from each other, so that both matrices can be expected to share similar properties. The next section is dedicated to further remarks on this underlying Laplace-like operator, which reflects the microscopic arrangements of individuals.

2.4 Underlying Laplace-like operator

We investigate here the properties of the matrix $BB^\star$, which was identified as a kind of discrete Laplace operator. We shall see that this matrix actually differs from the matrices that result from space discretization of elliptic problems, or more generally from discrete Laplace operators for electric networks.

Let us consider a configuration $\mathbf{q} \in K$ (like in Fig. 2), and the associated matrix B, each line of which expresses the constraint

$$-\mathbf{G}_{ij} \cdot \mathbf{u} \leq 0,$$

where $\mathbf{G}_{ij}$ is the gradient of the distance $D_{ij} = |\mathbf{q}_j - \mathbf{q}_i| - r_i - r_j$ with respect to the configuration vector $\mathbf{q} = (\mathbf{q}_1, \dots, \mathbf{q}_N)$. Let us start with some comment on the operator $B^\star$. As we already pointed out, it can be interpreted as a discrete gradient. Considering a set of Lagrange multipliers λ (i.e. interaction pressures between particles in contacts), $-B^\star$ assembles the corresponding force field acting on the particles. In case the configuration is structured (e.g. orthogonal lattice of discs of the same size, or the triangular lattice of Fig.. 3) then a constant pressure field induces no force at all (expect for the boundary of the cluster). It is the discrete counterpart of the standard property in the continuous setting that the gradient of

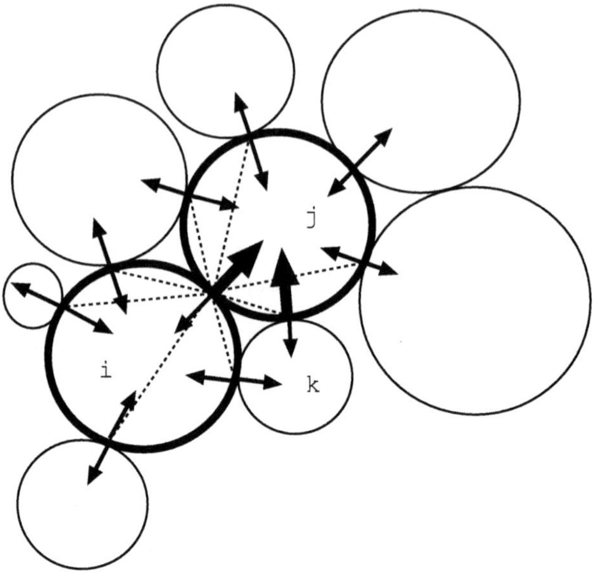

Figure 2. Non structured stencil

a constant field is zero. Yet, in the general situation (when no structural assumption is made on the local arrangements of discs), this property does not hold. See e.g. the discs on Fig. 2: the vectors pointing inward each particle do not sum up to zero. Another feature is typical of the unstructured discrete situation. Consider the cluster represented in Fig. 3. The number of discs is 14, thus the number of degrees of freedom is 28, whereas the number of contacts is 29. As a consequence, the kernel of $B^\star$ is not trivial: there exists a non-zero pressure field (one pressure for each contact point) such that the resulting force field is 0. A striking consequence of this fact is the following: the associated Laplace-like operator $BB^\star$ does *not* satisfy the Hopf maximum principle, since there exist pressure fields λ such that $BB^\star \geq 0$, whereas some pressures are negative.

The matrix $BB^\star$ can be expressed as follows: considering a pressure field $\lambda = (\lambda_{k\ell})$, where (k,ℓ) runs over the set of actual contacts, the vector $BB^\star\lambda$ is a pressure-like vector (one component for each contact), and the value corresponding to the contact between discs i and j is

$$\sum_{(k,\ell)\sim(i,j)} \lambda_{k\ell}\mathbf{G}_{ij} \cdot \mathbf{G}_{k\ell}.$$

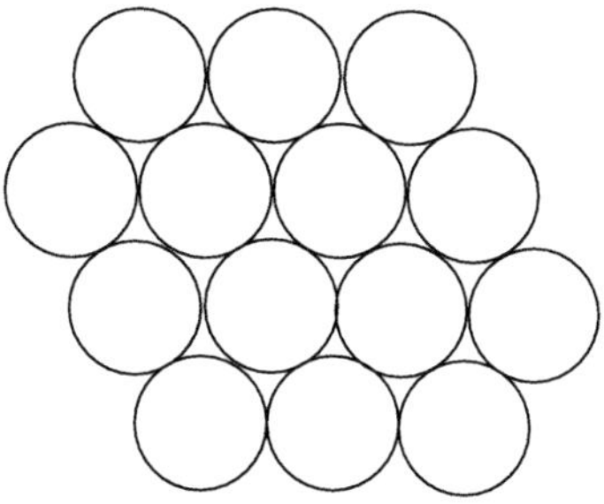

Figure 3. Hyperstatic situation

The discrete matrix operates on the network which is dual to the primary network (i.e. the network made of the centers of discs, with a connection between two vertices as soon as the discs are in contact). The vertices of the dual network are the contact points, and two vertices are connected, or equivalently the corresponding element of the matrix is non zero, as soon as the contacts share a same particle. Both networks are represented in Fig. 4. The corresponding *stencil* is represented in Fig. 2, around a vertex that is the contact point between two particles. The non-verification of the maximum principle is due to the fact that, when one considers three particles i, j, and k, it may happen that

$$\mathbf{e}_{ij} \cdot \mathbf{e}_{kj} > 0,$$

where $\mathbf{e}_{ij}$ is the unit vector $(\mathbf{q}_j - \mathbf{q}_i)/|\mathbf{q}_j - \mathbf{q}_i|$. Examples of such vectors are represented in Fig. 2, in bold. This property is generic in jammed collections of discs. As a consequence, some of the extra diagonal elements of $BB^\star$ are *positive*, thus $BB^\star$ is *not* a M-matrix. If one aims at interpreting the situation in terms of electric networks, in means that some resistances are *negative*, and this situation is native for jammed population of individual identified to rigid discs.

Condition number In terms of numerical computation, the inner difficulty of the problem can be quantified by the condition number of $BB^\star$ (i.e. the ratio between the largest and the smallest non zero eigenvalue of this symmetric positive matrix). This number κ can be related in some way to the lack of convexity of the feasible set K in the following sense: as detailed

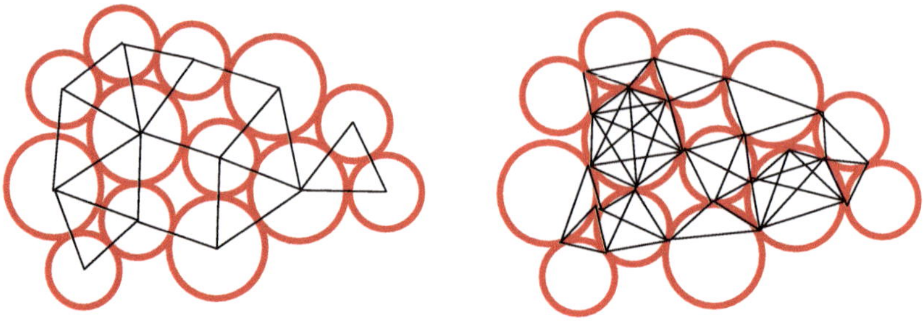

Figure 4. Primal (left) and dual (right) contact networks

in Maury and Venel (2011), it holds that

$$\kappa = \mathrm{cond}_2(BB^\star) \geq \frac{C}{\eta^2},$$

where η is the prox-regularity constant of K, i.e. the largest constant that can be chosen in Definition 2.3. This constant tends to zero very fast as the number of discs grows (see again Maury and Venel (2011)).

3 Macroscopic setting

3.1 The macroscopic model

We aim now at describing the population at a macroscopic level, i.e. by a density $\rho(x,t)$. We consider a bidimensional domain Ω (the room), we denote by $\mathbf{U}$ the desired velocity field (which is given), and by $\mathbf{u}$ the actual velocity field (which is affected by congestion effects). The set of feasible densities is defined as

$$K = \left\{\rho \in L^1(\Omega)\,,\ 0 \leq \rho(x) \leq 1 \quad a.e. \text{ in } \Omega\right\}.$$

The density is transported by the actual velocity field[4]:

$$\frac{\partial \rho}{\partial t} + \nabla \cdot (\rho \mathbf{u}) = 0, \tag{18}$$

where $\mathbf{u}$ is defined as the L^2 projection of $\mathbf{U}$ on the cone of feasible velocities $C_K(\rho)$. This model was proposed in Maury et al. (2010), and studied in the case where $\mathbf{U}$ is a gradient.

Unformally said, $C_K(\rho)$ contains velocity fields that do not increase the density where the constraint is already saturated (i.e. wherever $\rho = 1$). It can be defined in a dual way as

$$C_K(\rho) = \left\{\mathbf{v} \in L^2(\Omega)^2, \int_\Omega \mathbf{v} \cdot \nabla q \leq 0 \quad \forall q \in H^1_+(\Omega)\,, \int_\Omega q(1-\rho) = 0\right\},$$

$$\text{with } H^1_+(\Omega) = \{q \in H^1(\Omega)\,,\ q \geq 0 \text{ a.e. in } \Omega\}. \tag{19}$$

Note that the constraint on the integral of $q(1-\rho)$ imposes that q is 0 in the zones where the constraint is not saturated (i.e. $\rho < 1$).

When a part Σ of the boundary of Ω is an open exit, a free outlet condition can be prescribed by simply prescribing that the pressure is 0 on Σ (since the pressure is in $H^1(\Omega)$, its trace on Σ is indeed well-defined).

Saddle-point formulation The projection of the desired velocity on the cone of feasible ones can be formulated in a dual manner, in the form of a unilateral Darcy problem:

Find $(\mathbf{u}, p) \in L^2(\Omega)^2 \times H^1_\rho(\Omega)$, where

$$H^1_{\rho+}(\Omega) = \left\{q \in H^1_+(\Omega), \int_\Omega q(1-\rho) = 0\right\},$$

such that

$$\left\{\begin{array}{lcll} \mathbf{u} + \nabla p & = & \mathbf{U} & \\ -\displaystyle\int_\Omega \mathbf{u} \cdot \nabla q & \leq & 0 & \text{for all } q \in H^1_{\rho+}(\Omega), \end{array}\right. \tag{20}$$

with the complementarity condition:

$$-\int_\Omega \mathbf{u} \cdot \nabla p = 0\,.$$

Toward a suitable framework The model can be written

$$\frac{\partial \rho}{\partial t} + \nabla \cdot (\rho \mathbf{u}(\rho)) = 0,$$

[4]The transport equation is meant in a weak sense; we shall say that $\mathbf{u}$ transports ρ in the time interval $[0, T]$, with initial condition $\rho = \rho^0$, when

$$\int_0^T \int_\Omega \rho \partial_t \varphi + \int_0^T \int_\Omega \rho \mathbf{u} \cdot \nabla \varphi + \int_\Omega \varphi(0, \cdot)\rho^0 = 0 \quad \forall \varphi \in C_c^\infty([0, T) \times \Omega).$$

where the mapping $\rho \mapsto \mathbf{u}(\rho)$ corresponds to the projection of $\mathbf{U}$ on $C_K(\rho)$. This mapping is nonlinear, nonlocal, the dependence upon ρ is not smooth; it does not fit in the classical framework of conservation laws.

In the microscopic setting, the difference $\tilde{\mathbf{q}} - \mathbf{q}$ between the configurations $\mathbf{q}$ and $\tilde{\mathbf{q}}$ corresponds to a collection of individual displacements which reflects the *Lagrangian* character of the description. In the standard framework of Partial Differential Equations, the difference between entities (or, more generally, functions) $\tilde{\rho} - \rho$ is of different nature, because of the *Eulerian* character of the description. In order to extend the tools we used in the microscopic context, we must adopt another standpoint. Let us show that the Wasserstein distance, based on optimal transportation, provides an adapted framework.

3.2 Optimal transportation and Wasserstein distance

We present here some basics on optimal transportation, and we refer to Villani (1995) for a more detailed and more general presentation of those concepts. We assume here that the domain Ω is convex. For any measurable map $\mathbf{t} : \Omega \longrightarrow \Omega$, and probability densities[5] μ and ν supported in $\overline{\Omega}$, we say that ν is the *pushforward* of μ by $\mathbf{t}$ whenever

$$\int_{\mathbf{t}^{-1}(A)} \mu(x)\, dx = \int_A \nu(x)\, dx,$$

for any measurable set $A \subset \Omega$. Considering that the cost of moving x to y is $|y - x|^2$ (quadratic cost), the cost of the transport map $\mathbf{t}$ is defined as

$$C(\mathbf{t}) = \int_\Omega |\mathbf{t}(x) - x|^2\, dx.$$

The quadratic Wasserstein distance $W_2(\mu, \nu)$ is then defined by

$$W_2(\mu,\nu)^2 = \inf_{\mathbf{t},\, \mathbf{t}_\sharp \mu = \nu} C(\mathbf{t}) = \inf_{\mathbf{t},\, \mathbf{t}_\sharp \mu = \nu} \int_\Omega |\mathbf{t}(x) - x|^2\, dx.$$

In the case we considered (in particular the first measure is absolutely continuous), the minimizer is attained, and the minimization problem can be formalized in a dual way: it holds that

$$\frac{1}{2} W_2(\mu,\nu)^2 = \frac{1}{2} \min_{\mathbf{t},\, \mathbf{t}_\sharp \mu = \nu} \int_\Omega |\mathbf{t}(x) - x|^2\, dx \qquad (21)$$

[5] As detailed in Villani (1995), the approach generalizes to general measures, but we assume here absolute continuity, which is the case in the situation we consider.

$$= \max_{\varphi,\psi \in C_b(\Omega)} \left\{ \int_\Omega \varphi(x)\rho(x)\,dx + \int_\Omega \psi(y)\nu(y)\,dy\,,\ \varphi(x) + \psi(y) \le \frac{1}{2}\left|y - x\right|^2 \right\},$$

where $C_b(\Omega)$ is the space of all those functions that are bounded and continuous in Ω. The latter maximum is attained for a couple (φ, φ^c), where

$$\varphi^c(y) = \inf_x \left(\frac{1}{2}\left|x - y\right|^2 - \varphi(x) \right).$$

The function φ is called a *Kantorovich* potential for the transport problem from μ to ν; it is related to the transport map $\mathbf{t}$ that realizes the Wasserstein distance by

$$\mathbf{t} = \boldsymbol{i} - \nabla\varphi, \tag{22}$$

where $\boldsymbol{i}$ is the identity. Let us consider a toy problem to illustrate the considerations above. Let η be given in $(0, 1/2)$, and let I_η be the interval $(-1/2 - \eta/2, 1/2 + \eta/2) \subset \mathbb{R}$, the length of which is $1 - \eta$. We consider a probability density

$$\rho_\eta = \frac{1}{1-\eta}\mathbf{1}_{I_\eta},$$

(characteristic function of I_η, normalized to recover a unit mass). The density ρ_η violates the constraint and its projection on K is ρ_0, the characteristic function of the interval $(-1/2, 1/2)$. The Kantorovich potentials (from ρ_0 to ρ_η) can be computed exactly as

$$\varphi(x) = \frac{\eta}{2}x^2\,,\ \Psi(y) = -\frac{1}{2}\frac{1}{1-\eta}y^2.$$

The transport map from ρ_0 to ρ_η is indeed

$$x \longmapsto (\boldsymbol{i} - \nabla\varphi)\,x = (1-\eta)x,$$

and its inverse (ρ_η to ρ_0) is

$$y \longmapsto (\boldsymbol{i} - \nabla\psi)\,y = y + \frac{\eta}{1-\eta}y = \frac{1}{1-\eta}y.$$

3.3 Catching-up algorithm

Let $\tau > 0$ be a time step. The catching-up algorithm consists in transporting the current density by the desired velocity field, and then projecting it on the set K of feasible densities:

$$\tilde\rho^{n+1} = (\boldsymbol{i} + \tau\mathbf{U})_{\#}\,\rho^n \quad \text{(prediction step)}, \tag{23}$$

$$\rho^{n+1} = P_K \tilde{\rho}^{n+1} \quad \text{(correction step)}, \tag{24}$$

where the projection is performed in the Wasserstein sense. This projection can be shown to be well defined[6] (Maury et al., 2011).

In the microscopic setting, the link between the catching-up algorithm and the crowd motion model was straightforward[7]. In the present situation, the link is less straightforward, and the proof of convergence of the discrete trajectories requires technical developments that we will not describe here (see Maury et al. (2010, 2011)). We shall simply describe here the core of the proof, which lies in the link between the catching up scheme and the unilateral Darcy problem (20).

In the microscopic setting, the actual velocity at the discrete level was written $(\mathbf{q}^{n+1} - \mathbf{q}^n)/\tau$ (see e.g. (8)). This expression has to replaced by a similar expression involving ρ^n and

$$\rho^{n+1} = P_K\left((\boldsymbol{i} + \tau\mathbf{U})_{\#}\, \rho^n\right),$$

which would be a Lagrangian version of "$(\rho^{n+1} - \rho^n)/\tau$", so to say. The

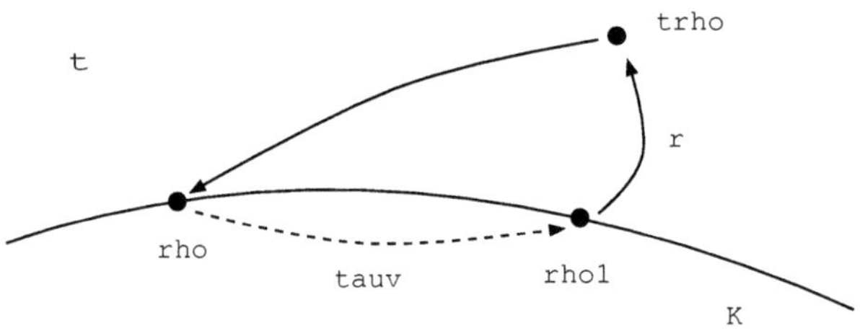

Figure 5. Definition of the discrete transport maps.

situation is represented in Fig. 5. The main idea consists in defining a

[6]This fact is somewhat striking since, in the microscopic setting, the set K can be shown to be η-prox-regular with η going to 0 as the number of individuals goes to ∞, and their size goes to 0. It means that the projection is defined in a neighborhood of ∂K which shrinks down to ∂K itself as the population grows, which suggests a degenerated behavior when it tends to the macroscopic situation. The very fact that it does not degenerate in the macroscopic situation (the projection is well-defined, no matter what the distance to K is) reflects the deep difference between the two approaches, as it was already addressed at the end of Section 1.

[7]We refer the reader to the few lines between Eq. (8) (which is a discretization of (5)) and Eq. (9)

discrete velocity $\mathbf{u}^{n+1}$ such that $\tau \mathbf{u}^{n+1}$ corresponds to the displacement between ρ^n and ρ^{n+1}, and to show that this velocity solves (at least at the first order in time) a Darcy problem like (20). It can be done as follows: we define $\mathbf{t^{n+1}}$ as $(\boldsymbol{i}+\tau\mathbf{U})^{-1}$, which is well defined as soon as $\mathbf{U}$ is regular and τ is small enough, and $\mathbf{r}^{n+1}$ as the optimal map between ρ^{n+1} and $\tilde{\rho}^{n+1}$. The discrete velocity is then defined as

$$\mathbf{u}^{n+1} = \frac{\boldsymbol{i} - \mathbf{t^{n+1}} \circ \mathbf{r}^{n+1}}{\tau}.$$

Notice that this velocity is defined in the target set: considering a element of mass x of ρ^n, transported to y in ρ^{n+1}, it holds

$$y = x + \tau \mathbf{u}^{n+1}(y),$$

where the velocity is defined at y, and not at x.

Let us introduce

$$\mathbf{w}^{n+1} = \frac{\boldsymbol{i} - \mathbf{r}^{n+1}}{\tau} \Longleftrightarrow \mathbf{r}^{n+1} = \boldsymbol{i} - \tau \mathbf{w}^{n+1}.$$

As soon as $\mathbf{U}$ is assumed to be smooth (i.e. continuously differentiable), $\mathbf{t}^{n+1}$ can be expanded at the first order as

$$\mathbf{t}^{n+1} = (\boldsymbol{i} + \tau\mathbf{U})^{-1} = \boldsymbol{i} - \tau\mathbf{U} + o(\tau),$$

where the $o(\tau)$ if uniform with respect to the space variable x. We then have

$$\begin{aligned}\mathbf{u}^{n+1} &= \frac{1}{\tau}\left(\boldsymbol{i} - (\boldsymbol{i} - \tau\mathbf{U} + o(\tau)) \circ (\boldsymbol{i} - \tau\mathbf{w}^{n+1})\right)\\ &= \mathbf{w}^{n+1} + \mathbf{U} + O(\tau). \qquad (25)\end{aligned}$$

To complete the identification with (20), we still have to establish that $\mathbf{w}^{n+1}$ is the gradient of a pressure p that is nonnegative, and that vanishes outside the saturated zone. To that purpose, we follow the strategy introduced in Butazzo and Santambrogio (2005) and used in the context of crowd motion models of the gradient flow type in Maury et al. (2010). It consists in proving that the Kantorovich potential associated to the transport problem from ρ^{n+1} to $\tilde{\rho}^{n+1}$ (which corresponds to the displacement $-\tau\mathbf{w}^{n+1}$) can be interpreted (up to multiplicative and additive constants) as a pressure field in the saturated zone. We shall assume here that ρ^{n+1} is positive in Ω (we refer to Maury et al. (2010) for an adaptation to the general case). The proof is somewhat paradoxical, since it consists in considering *eulerian* variations around the minimizer, whereas the Wasserstein framework suggests to build variations by means of transport maps (i.e. *horizontal* variations). Consider μ a density in K, and $\varepsilon > 0$. We define

$$\rho_\varepsilon = \rho^{n+1} + \varepsilon(\mu - \rho^{n+1}) \in K,$$

and we denote by φ_ε and ψ_ε the Kantorovich potentials associated to the transport problem from ρ_ε to $\tilde{\rho}^{n+1}$. By the Monge Kantorovich formulation (21), it holds that

$$\frac{1}{2} W_2(\rho_\varepsilon, \tilde{\rho}^{n+1})^2 = \int_\Omega \varphi_\varepsilon \rho_\varepsilon + \int_\Omega \psi_\varepsilon \tilde{\rho}^{n+1}, \tag{26}$$

and

$$\frac{1}{2} W_2(\rho^{n+1}, \tilde{\rho}^{n+1})^2 =$$

$$\max_{\varphi, \psi \in C_b(\Omega)} \left\{ \int_\Omega \varphi \rho^{n+1} + \int_\Omega \psi \tilde{\rho}^{n+1}, \ \varphi(x) + \psi(y) \leq \frac{1}{2} |y - x|^2 \right\},$$

$$\geq \int_\Omega \varphi_\varepsilon \rho^{n+1} + \int_\Omega \psi_\varepsilon \tilde{\rho}^{n+1}. \tag{27}$$

Besides, since ρ^{n+1} minimizes the Wassertein distance from $\tilde{\rho}^{n+1}$ to K by construction, and since ρ_ε is in K, we have

$$\frac{1}{2} W_2(\rho_\varepsilon, \tilde{\rho}^{n+1})^2 \geq \frac{1}{2} W_2(\rho^{n+1}, \tilde{\rho}^{n+1})^2,$$

which yields, thanks to (26) and (27),

$$\int_\Omega \varphi_\varepsilon(\rho_\varepsilon - \rho^{n+1}) \geq 0 \Longrightarrow \int_\Omega \varphi_\varepsilon \rho^{n+1} \leq \int_\Omega \varphi_\varepsilon \mu \quad \forall \mu \in K. \tag{28}$$

Now, prescribing a fixed value of the Kantorovich potentials at some point $x_0 \in \Omega$, the Kantorovich potential is unique, and the sequence (φ_ε) can be shown (see Butazzo and Santambrogio (2005), Lemma 3.4) to converge to the Kantorovich potential φ associated to the transport from ρ^{n+1} to $\tilde{\rho}^{n+1}$ (i.e. in the limit $\varepsilon = 0$). So finally ρ^{n+1} minimizes a linear functional of the type

$$\rho \longmapsto \int_\Omega \varphi \rho,$$

over K which is the set of probability densities that are less that 1 almost everywhere. If the upper bound constraint were not prescribed, ρ would tend to concentrate on the minimizer(s) of φ. This concentration is ruled

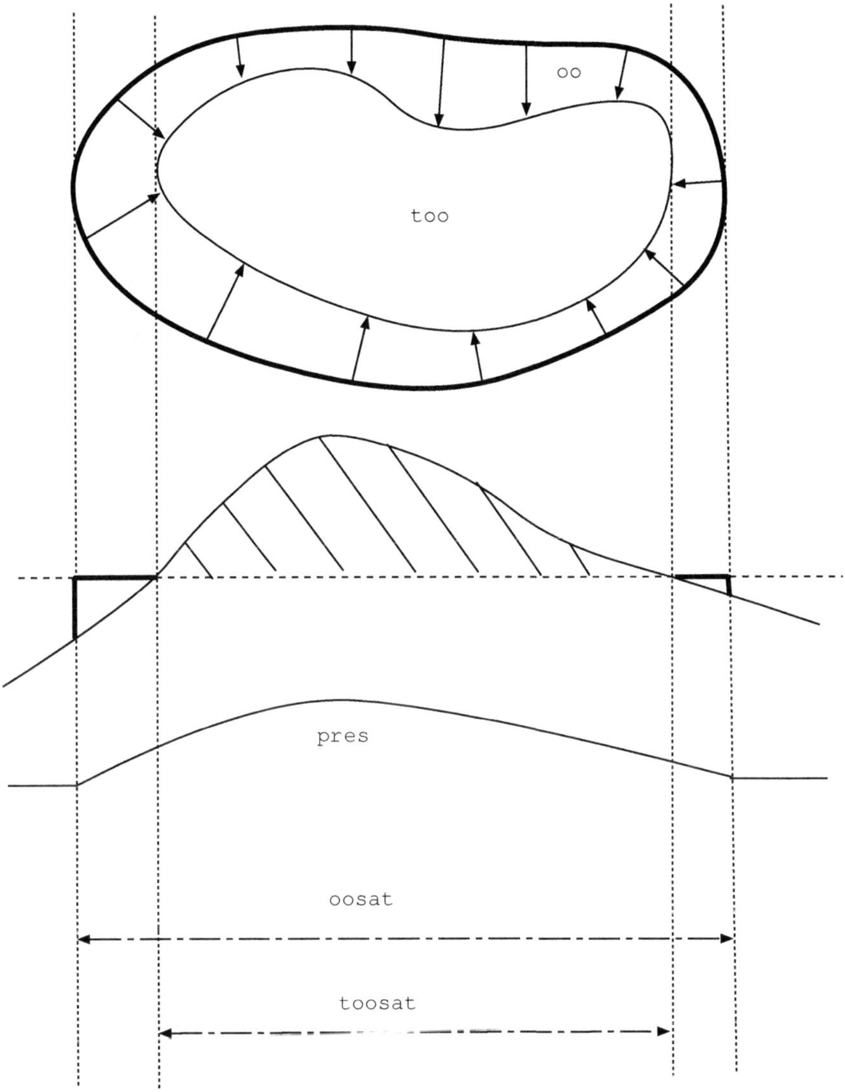

Figure 6. Projection of $\tilde{\rho}$ onto K

out by the congestion constraint, and ρ saturates the zone in which φ is minimal, so that it takes the following form

$$\rho^{n+1} \left| \begin{array}{lll} = 1 & \text{on} & [\varphi < \ell] \\ \leq 1 & \text{on} & [\varphi = \ell] \\ = 0 & \text{on} & [\varphi > \ell] \end{array} \right.$$

where ℓ is a value which is adjusted to comply with the unit mass constraint. In the previous expression, the way ρ^{n+1} is distributed over $[\varphi = \ell]$ is underdetermined, since it does not affect the value of the functional: the solution of the previous minimization problem is not unique. It does not matter here, since ρ^{n+1} was already identified as the unique density that realizes the distance to K. We then define the pressure field

$$p = \frac{1}{\tau}\left(\ell - \varphi\right)^{+}.$$

By construction, p is nonnegative, it vanishes outside of the saturated zone (i.e. where $\rho^{n+1} < 1$), and $\nabla p = -\nabla\varphi/\tau$ in the support of p. Now recall that φ is a Kantorovich potential for $\mathbf{r}^{n+1}$, i.e.

$$\mathbf{r}^{n+1} = \boldsymbol{i} - \nabla\varphi = \boldsymbol{i} + \tau\nabla p.$$

Finally, we have

$$\mathbf{w}^{n+1} = \frac{\boldsymbol{i} - \mathbf{r}^{n+1}}{\tau} = -\nabla p,$$

so that, thanks to (25), we obtain the Darcy decomposition (at the first order in time) of the desired velocity field $\mathbf{U}$ as the sum of the actual velocity and the gradient of a pressure:

$$\mathbf{u}^{n+1} + \nabla p = \mathbf{U} + O(\tau).$$

Fig. 6 is an attempt to illustrate this construction: the zone in which the constraint is violated (i.e. $\tilde{\rho}^{n+1} > 1$) is denoted by $\tilde{\omega}$. The excess of mass is spread out around this zone, and the obtained density saturates the constraint ($\rho^{n+1} \equiv 1$) on ω. Note again that the displacement $\tau\mathbf{u}^{n+1}$ which pushes ρ^n onto ρ^{n+1} is defined *on the target set*, which rules out a straight use of this approach to design a tractable numerical procedure.

3.4 Numerical issues

Like in the microscopic setting, a direct use of the saddle point form of the projection onto K is delicate, since the pressure field is defined on the target set. We describe here a Monte Carlo algorithm based on a stochastic interpretation of the Poisson problem with Dirichlet boundary condition (see Maury et al. (2011) for details). Let us make it clear that this approach is not covered by a rigorous numerical analysis. We consider again the situation represented in Fig. 5, and we drop the upperscripts $^{n+1}$ to alleviate notation. The predicted density $\tilde{\rho}$ violates the constraint in the domain $\tilde{\omega}$, and the excess mass is $(\tilde{\rho} - 1)^+$. Since this excess mass is due to the transport of an admissible density by a contracting field (desired velocity field $\mathbf{U}$) during τ, it is of the order τ. Let us assume that it can be written $\tau\nu$, where ν is a nonnegative density supported by the oversaturated zone[8].

The density $\tilde{\rho}$ is $1 + \tau\nu$ in $\tilde{\omega}$. Let ρ be the projection of $\tilde{\rho}$ on K. The displacement from ρ to $\tilde{\rho}$ is of the form $\mathbf{r} = \boldsymbol{i} + \tau\nabla p$, where p is 0 outside ω (see the previous section). Since ρ saturates the constraint[9] in ω, we have in ω

$$\frac{1}{|\boldsymbol{i} + \tau\nabla p|}(x) \approx (1 - \tau\nabla \cdot \nabla p)\,(x) = 1 + \tau\nu(\mathbf{r}(x)),$$

so that, at the first order in τ,

$$-\Delta p(x) = \nu(\mathbf{r}(x)),$$

with $p = 0$ on the boundary of ω. The unknown pressure field p and the right-hand side are not defined on the same set, but $\mathbf{r}$ is close to the identity $\boldsymbol{i}$. Replacing $\nu(\mathbf{r}(x))$ by $\nu(x)$ is audacious, even in this informal approach, since ν has no reason to be regular. Yet, ν and $\nu \circ \mathbf{r}$ are close (of the order τ) in the Wasserstein sense by construction, therefore their distance in the H^{-1} sense is of the same order (see Maury et al. (2010), Lemma 3.4). Thus, the corresponding pressures are also close (at the first order in τ) in the H^1 norm. Using again the fact that ω is transported to $\tilde{\omega}$ by $\mathbf{r}$, which is the identity at the first order in τ, and assuming sufficient regularity to transport the elliptic problem back to $\tilde{\omega}$ (up to first order terms), we end up with a Poisson problem in $\tilde{\omega}$

$$-\Delta p = \nu \quad \text{in } \tilde{\omega},$$

with homogeneous Dirichlet boundary conditions on $\partial\tilde{\omega}$.

The velocity of the boundary is $-\partial p/\partial n$, which means that the quantity of mass crossing an element $d\gamma$ of the boundary during τ is $-\tau\,\partial p/\partial n\, d\gamma$. Now consider the stochastic interpretation of the Poisson problem (we assume that ν has unit mass, which can be recovered by a straightforward

[8]Since $\tilde{\rho}$ comes from the transport of the previous density ρ^n by $\mathbf{U}$, it holds approximately $\tilde{\rho} \approx 1 - \tau\nabla \cdot \mathbf{U} + o(\tau)$.

[9]This property is straightforward: if it were not, it would obviously not be the density closest to $\tilde{\rho}$ in K.

renormalization): consider a brownian motion starting at an initial point X, which is supposed to be itself a random position following the law of density ν in $\tilde{\omega}$, and consider the location at which this Brownian motion crosses the boundary of $\tilde{\omega}$. The law of the position of this first hit is known to follow a law with density $-\partial p/\partial n$ on $\partial\tilde{\omega}$, where p is the solution to the Poisson problem above. The algorithm we propose (see Maury et al. (2011)) is deduced from these consideration.

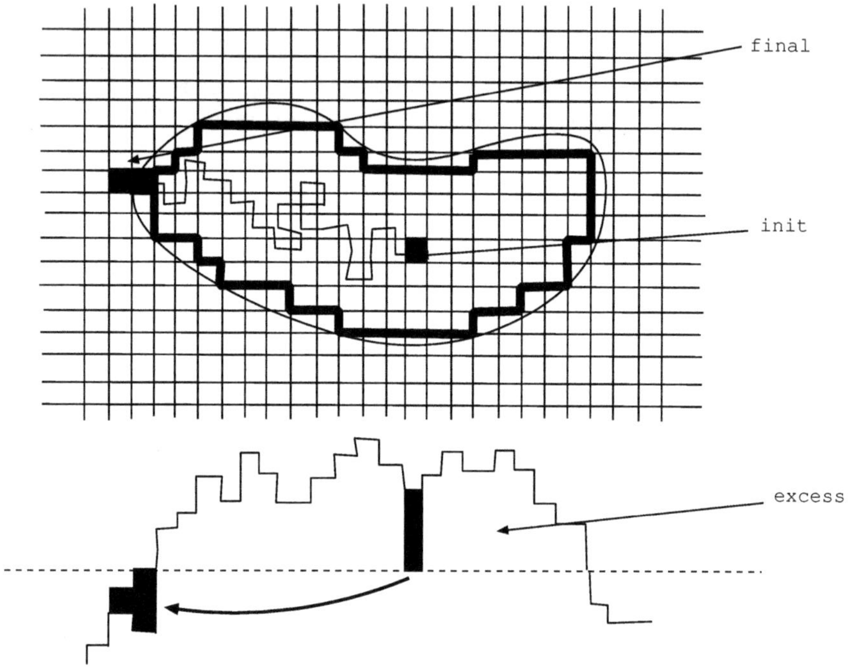

Figure 7. Random walk algorithm

The space is discretized in a finite volume spirit by a cartesian mesh. The densities are assumed to be constant on each cell of this mesh. We illustrate in Fig. 7 the discretized counterpart of the situation represented in Fig. 6. The zone delimited by the bold broken line corresponds to cells where the constraint is violated. Considering one of those cells as a starting point, with an excess of mass m, we run a random walk on the grid with balanced transition probabilities. When the random walk reaches a cell which is not saturated, the excess of mass is put in this cell, up to saturation. If there is not enough empty space to get rid of the total excess m, the random walk continues according to the same principle, until there is no mass left. Fig. 7 illustrates the process, starting from a cell inside the zone where the constraint is violated.

A new random walk is then initiated from another cell where the constraint is saturated, and so on, until the constraint is satisfied everywhere. We refer to Maury et al. (2011) for a more detailed description of the algorithm, and for some illustrations of the behavior of the algorithm.

Bibliography

H. H. Bauschke and P. L. Combettes. *Convex analysis and monotone operator theory in Hilbert spaces.* CMS Books in Mathematics/Ouvrages de Mathématiques de la SMC. Springer, New York, 2011.

G. Butazzo and F. Santambrogio. A model for the optimal planning of an urban area. *SIAM J. Math. Anal.*, pages 514–530, 2005.

P. G. Ciarlet. *Introduction to Numerical Linear Algebra and Optimisation.* Cambridge Texts in Applied Mathematics. Cambridge University Press, Cambridge, 1989.

F. H. Clarke, R. J. Stern, and P. R. Wolenski. Proximal smoothness and the lower-C2 property. *J. Convex Anal.*, pages 117–144, 1995.

B. Maury. A time-stepping scheme for inelastic collisions. numerical handling of the nonoverlapping constraint. *Numerische Mathematik*, 102: 649–679, 2006.

B. Maury and J. Venel. A discrete contact model for crowd motion. *ESAIM Mathematical Modelling and Numerical Analysis*, 45:145–168, 2011.

B. Maury, A. Roudneff-Chupin, and F. Santambrogio. A macroscopic crowd motion model of gradient flow type. *Mathematical Models and Methods in Applied Sciences*, 20:1787–1821, 2010.

B. Maury, A. Roudneff-Chupin, F. Santambrogio, and J. Venel. A macroscopic crowd motion model of gradient flow type. *Networks and Heterogeneous Media*, 6(3):485–519, September 2011.

J.-J. Moreau. Décomposition orthogonale d'un espace Hilbertien selon deux cônes mutuellement polaires. *C. R. Acad. Sci. Paris*, 255:238–240, 1962.

J.-J. Moreau. Evolution problem associated with a moving convex set in a hilbert space. *J. Differential Equations*, 26:346–374, 1977.

R. A. Poliquin and R. T. Rockafellar. Prox-regular functions in variational analysis. *Trans. Amer. Math. Soc.*, 348:1805–1838, 1996.

R.T. Rockafellar. *Convex Analysis.* Princeton mathematical series. Princeton University Press, 1970.

J. Venel. A numerical scheme for a class of sweeping processes. *Numerische Mathematik*, 118:367–400, 2011.

C. Villani. *Topics in Optimal Transportation*, volume 58. AMS, 1995.

Pedestrians moving in the dark: Balancing measures and playing games on lattices

Adrian Muntean*† and Emilio N. M. Cirillo‡
Oleh Krehel* and Michael Böhm**

* CASA - Center for Analysis, Scientific computing and Applications, Department of Mathematics and Computer Science, Eindhoven University of Technology, The Netherlands

† Institute for Complex Molecular Systems (ICMS), Eindhoven University of Technology, The Netherlands

‡ Dipartimento di Scienze di Base e Applicate per l'Ingegneria, Sapienza Università di Roma, Italy

** Zentrum für Technomathematik, Fachbereich Mathematik und Informatik, Universität Bremen, Germany

Abstract We present two conceptually new modeling approaches aimed at describing the motion of pedestrians in obscured corridors:

(i) a Becker-Döring-type dynamics and
(ii) a probabilistic cellular automaton model.

In both models the group formation is affected by a threshold. The pedestrians are supposed to have very limited knowledge about their current position and their neighborhood; they can form groups up to a certain size and they can leave them. Their main goal is to find the exit of the corridor.

Although being of mathematically different character, the discussion of both models shows that it seems to be a disadvantage for the individual to adhere to larger groups.

We illustrate this effect numerically by solving both model systems. Finally we list some of our main open questions and conjectures.

1 Introduction

Social mechanics is a topic that has attracted the attention of researchers for more than one hundred years; see e.g. (Haret, 1910; Portuondo y Barceló, 1912). A large variety of existing models are able to describe the dynamics of pedestrians driven by a *desired velocity* towards clearly defined exits. But how can we possibly describe the motion of pedestrians when the exits are not clearly defined, or even worse, *what if the exits are not visible*?

This paper is inspired by a practical evacuation scenario. Some of the existing models are geared towards describing the dynamics of pedestrians with somehow given, prescribed or, at least, desired velocities or spatial fluxes towards an exit the location of which is, more or less, known to the pedestrians[1]. We focus on modeling basic features which we assume to be influencing the motion of pedestrians in regions with reduced or no visibility[2]. Our scenario is the following: A large number of pedestrians, generally denoted by Y, is supposed to move through an obscured corridor,Ω. Due to the lack of visibility (e.g. smoke, fog, darkness, etc.[3]) the Y's cannot see the exit. We allow for some sort of "buddying": If Y's hit each other they might decide to form a group. For practical reasons, we limit the size of such groups by a threshold T. As transport mechanism, we assume a very mild diffusion-like motion which is not connected with the location of the exit. To model this situation, we take two different routes by introducing and discussing:

(1) a Becker-Döring-type system of balance equations for mass measures (see Appendix A for a derivation)

(2) a lattice model for an interacting particle system with threshold dynamics.

The two approaches are conceptually different. They consider from two different perspectives the concept of *group* (social collectivity). In the following sections, we approximate the corresponding dynamics for evacuation scenarios similar to those described in Fang et al. (2012) and Zheng et al. (2011), for instance. In the first approach, the *group feature* is imbedded in a size-dependent mass measure and the evolution will be dictated by the con-

[1]Efficient evacuation of humans from high–risk zones is a very important issue cf. Schadschneider et al. (2009). The topic is very well studied by large communities of scientists ranging from logistics and transportation, civil and fire engineering, to theoretical physics and applied mathematics. Models (deterministic or stochastic) succeed to capture basic behaviors of humans walking within given geometries towards *a priori* prescribed exits. Typical classes of crowd dynamics models include social force/social velocity models (cf. e.g. Helbing and Molnar (1995), Piccoli and Tosin (2011), Evers and Muntean (2011)), simple asymmetric exclusion models (see chapters 3 and 4 from Schadschneider et al. (2011) as well as references cited therein), cellular automaton-type models Kirchner and Schadschneider (2002); Guo et al. (2012), etc.; a detailed classification of pedestrian models, see Schadschneider et al. (2011), e.g.

[2]In recent years, high-rise buildings claim steadily increasing numbers of victims in evacuations. Most victims were due to the reduced visibility by fire smoke; see Jin (1978); Jin and Yamada (1985). In the future, most likely one will insist also on building underground, so the potential of smoke victims further increases. We refer the reader to Kobes et al. (2010) for a recent literature review.

[3]Think about an evacuation in a metro in which there is smoke and/or no light, etc.

servation equation of the respective measure (balancing the size-dependent density). In the second approach, we use a threshold to allow finite non-exclusion per site in a lattice automaton for the self-propelled particles (i.e. the pedestrians). We suspect however that connections between (1) and (2) might exist, but we don't expect that the mean-field limit of (2) is (1) (cf. e.g. Presutti (2013)).

Whatever route we take, our central questions are:

(Q1) *How do pedestrians choose their path and speed when they are about to move through regions with no visibility?*

(Q2) *Is group formation (e.g. buddying) the right strategy to move through such uncomfortable zones able to ensure exiting within a reasonable time?*

Answers to (Q1) and (Q2) are largely unknown. Group psychology (compare e.g. Le Bon (2008); Curşeu (2009) and Dyer et al. (2009)) lacks extensive experimental observations, and, due to absence of meaningful statistics, nothing can be really concluded. The "groups" we study here are expected to be highly unstable and therefore they only remotely resemble the well-studied swarming patterns typically observed in nature by fish and or birds communities (see e.g. the 4–groups taxonomy in Topaz and Bertozzi (2004), namely swarm, torus, dynamic parallel groups, and highly parallel groups).

The basic idea is the following: In the situation we are modeling, neighbors (both individuals or groups) can not be visually identified by the individuals in motion, so that basic mechanisms like attraction to a group, tendency to align, or social repulsion are negligible and individuals have to live with "preferences".

The paper is structured as follows: We start off with a continuum model describing the mesoscopic dynamics of groups in Section 2. After giving the set of governing equations in Section 2.1, we illustrate numerically the observed threshold effects at such mesoscopic level in Section 2.2. Appendix A contains a formal derivation in terms of mass measures of the Becker-Döring-like system proposed here. As next step, we propose a lattice model to capture the microscopic dynamics, see Section 3. The model detailed in Section 3.1 is illustrated numerically in Section 3.2. We conclude by enumerating a set of basic questions that are for the moment open (see Section 4) on the behavior of both interacting particle systems and structured densities with threshold effects.

2 Becker-Döring grouping in action

2.1 From interacting colloids to group dynamics

Inspired by the modeling of charged colloids transport in porous media (see e.g. Krehel et al. (2012); Ray et al. (2012)), we consider now a system of reaction-diffusion equations describing the aggregation and dissolution of groups; the ith variable in the vector of unknowns represents the specific size of the subgroup i (density of the i-mer u_i). Here u_1 – density of crowds of group size one (individuals), u_2 – density of groups of size two, and so on until u_N are the corresponding Radon-Nikodym derivatives of suitable measures (see Appendix A for details). For convenience, we take here $T := N$, the biggest group size.

The following equations describe our system:

$$\partial_t u_1 + \nabla \cdot (-d_1 \nabla u_1) = -u_1 \sum_{i=1}^{N-1} \beta_i u_i + \sum_{i=2}^{N} \alpha_i u_i - \beta_1 u_1 u_1 + \alpha_2 u_2 \quad (1)$$

$$\partial_t u_2 + \nabla \cdot (-d_2 \nabla u_2) = \beta_1 u_1 u_1 - \beta_2 u_2 u_1 + \alpha_3 u_3 - \alpha_2 u_2 \quad (2)$$

$$\vdots \quad (3)$$

$$\partial_t u_{N-1} + \nabla \cdot (-d_{N-1} \nabla u_{N-1}) = \beta_{N-2} u_{N-2} u_1 - \quad (4)$$

$$-\beta_{N-1} u_{N-1} u_1 + \alpha_N u_N - \alpha_{N-1} u_{N-1} \quad (5)$$

$$\partial_t u_N + \nabla \cdot (-d_N \nabla u_N) = \beta_{N-1} u_{N-1} u_1 - \alpha_N u_N. \quad (6)$$

This system of partial differential equations indicates that groups diffuse inside Ω. If the groups meet each other, then they start to interact via the mechanism suggested by the right-hand side of the system (aggregation or degradation being the only allowed interaction behaviors). We take as boundary conditions

$$u_1 = 0 \qquad \text{on } \Gamma_D \quad (7)$$

$$-d_1 \nabla u_1 \cdot n = 0 \qquad \text{on } \partial\Omega \setminus \Gamma_D \quad (8)$$

$$-d_i \nabla u_i \cdot n = 0 \qquad \text{on } \partial\Omega, i \in \{2, \dots, N\}, \quad (9)$$

while the initial conditions at $t = 0$ are

$$u_1 = M \qquad \text{in } \Omega \quad (10)$$

$$u_i = 0 \qquad \text{in } \Omega, i \in \{2, \dots, N\}. \quad (11)$$

These boundary conditions model the following scenario: Only the population of size one are allowed to exit, all the other groups need to split in smaller groups close to Γ_D. In (10), $M > 0$ denotes the initial density of individuals, the total mass [of pedestrians] in the system being $\int_\Omega \sum_{i=1}^{N} i u_i$.

The total mass at $t = 0$ is $M|\Omega|$. Note that (10) indicates that, initially, groups are not yet formed. Group formation happens here immediately after the initial time. As transport mechanism, we have chosen to use Fickian diffusion fluxes to model the mesoscopic erratic motion of the crowd [with all its N group structures] inside the corridor Ω.

Similarly to the case of moving colloidal particles in porous media (cf. for instance Krehel et al. (2012) and references cited therein), we take as reference diffusion coefficients the ones given the Stokes-Einstein relation, i.e. the diffusion coefficient of the social conglomeration is inversely proportional to its size as described by $d_i := \frac{1}{\sqrt[3]{i}}$ (which would correspond to the colloidal particles diffusion in a 3D confinement) for any $i \in \{1, \ldots, N\}$; see for instance Edward (1970). In contrast to the case of transport in porous media, we assume that no heterogeneities are present inside Ω. Consequently, the diffusion coefficients are taken here to be independent of the space and time variables. If heterogeneities were present (like it is nearly always the case e.g. in shopping malls), then one needs to introduce concepts like local porosity and porosity measures as in Evers and Muntean (2011); see Chepizhko et al. (2013) for a related scenario discussing stochastically interacting self propelled particles within a heterogeneous media with dynamic obstacles. We restrict ourselves here to the case of homogeneous corridors.

We take the degradation (dissociation, group splitting) coefficients $\alpha_i > 0$ ($i \in \{2, \ldots, N\}$) as being given constants, while for the aggregation coefficients we use the concept of *social threshold*. We define

$$\beta_i := \begin{cases} i & i < T \\ 1 & \text{otherwise}, \end{cases} \tag{12}$$

where $T \in (0, \infty)$ is the social threshold. Essentially, using (12) we expect that the choice of T essentially limits the size of groups that can be formed by means of this Becker-Döring-like model. In other words, even if large values of N are allowed (say mimicking $N \to \infty$) most likely groups of sizes around $\lfloor T \rfloor$ will be created; here $\lfloor p \rfloor$ denotes the integer part of $p \in \mathbb{R}$.

2.2 Threshold effects on mesoscopic group formation

For the numerical examples illustrated here, we consider $N = 20$ species waking inside the corridor $\Omega = (0, 1) \times (0, 1)$. On the boundary $\partial\Omega$, we design the door $\Gamma_D = \{(x, y) : x = 0, y \in [0.4, 0.6]\}$, while the rest of the boundary $\partial\Omega \setminus \Gamma_D$ is considered to be impermeable, i.e. the pedestrians cannot penetrate the wall $\partial\Omega \setminus \Gamma_D$.

To solve the system numerically, we use the library DUNE and rely on a 2D Finite Element method discretization (with linear Lagrange elements) for the space variable, with implicit time-stepping. Note that we allow only crowds of size one, i.e. u_1, to exit the door. For larger group sizes the door in impenetrable. Such groups really need to dissociate/degrade first and then attempt to exit. We choose constant degradation coefficients and take as reference values $\alpha_i = 0.7$ ($i \in \{1, \ldots, N\}$).

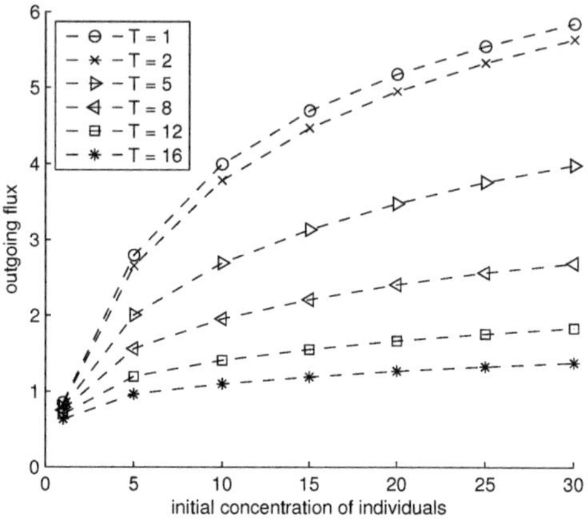

Figure 1. Outgoing flux with respect to initial density.

As we can see from Figure 1, the outgoing flux (close to the steady state[4]) exhibits a polynomial behavior with respect to the initial mass, where the polynomial exponent is influenced by the choice of the threshold T. It seems that the higher the threshold, the smaller is the polynomial power. This effect is rather dramatic – it indicates that, regardless the threshold size, behaving/moving gregariously is less efficient that performing random walks.

Figure 2 shows that there's no apparent saturation for the outgoing flux with respect to the mass: the growth goes on in a polynomial fashion. The linear behavior has been obtained by setting to zero the aggregation and degradation coefficients.

In Figure 3, we see that the influence of variable diffusion coefficients is marginal; since a lot of mass exchange is happening in terms of species u_1, setting all the other coefficients $d_2, \ldots, d_N$ to be lower than $d_1 = 1$ (i.e. bigger groups move somewhat slower than individuals) does not affect the

[4]The mass exiting the system is evenly distribute throughout the domain Ω.

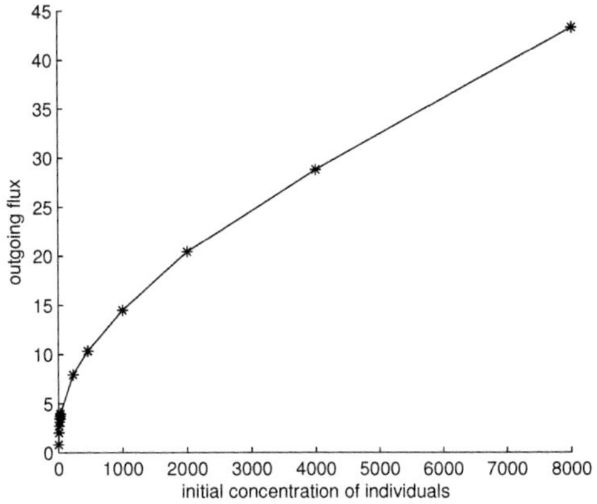

Figure 2. Outgoing flux for $T = 5$ *versus* large initial data $M \to \infty$.

output too much. Probably, the effect of diffusion could be stronger as soon as the effective diffusion coefficients are allowed to degenerate with locally vanishing u_i; this is a situation that can be foreseen in a modified setting Guo et al. (1988).

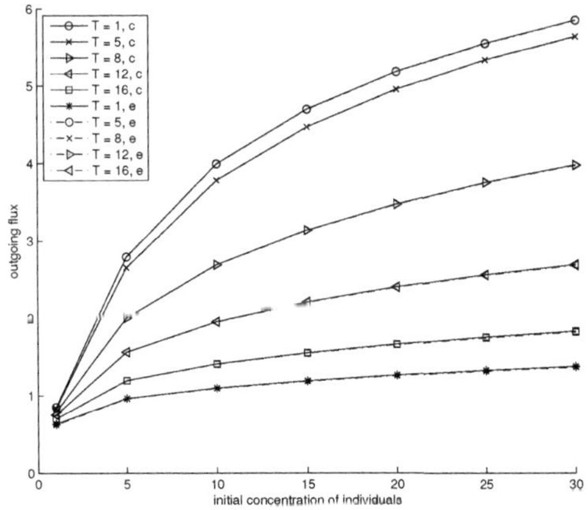

Figure 3. Homogeneous diffusion(c) and Stokes-Einstein diffusion(e). Note that the profiles are overlapping very closely.

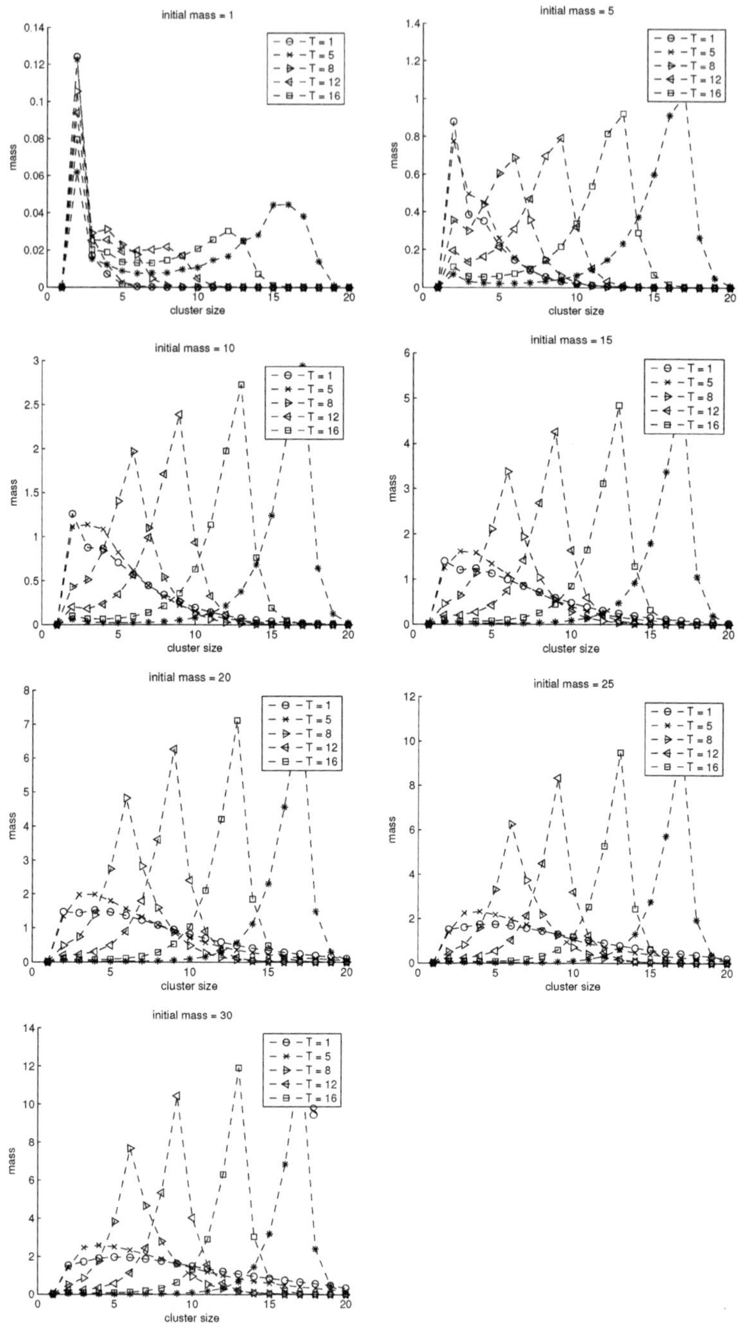

Figure 4. Steady-state mass distributions. Pile-up effect around group size T.

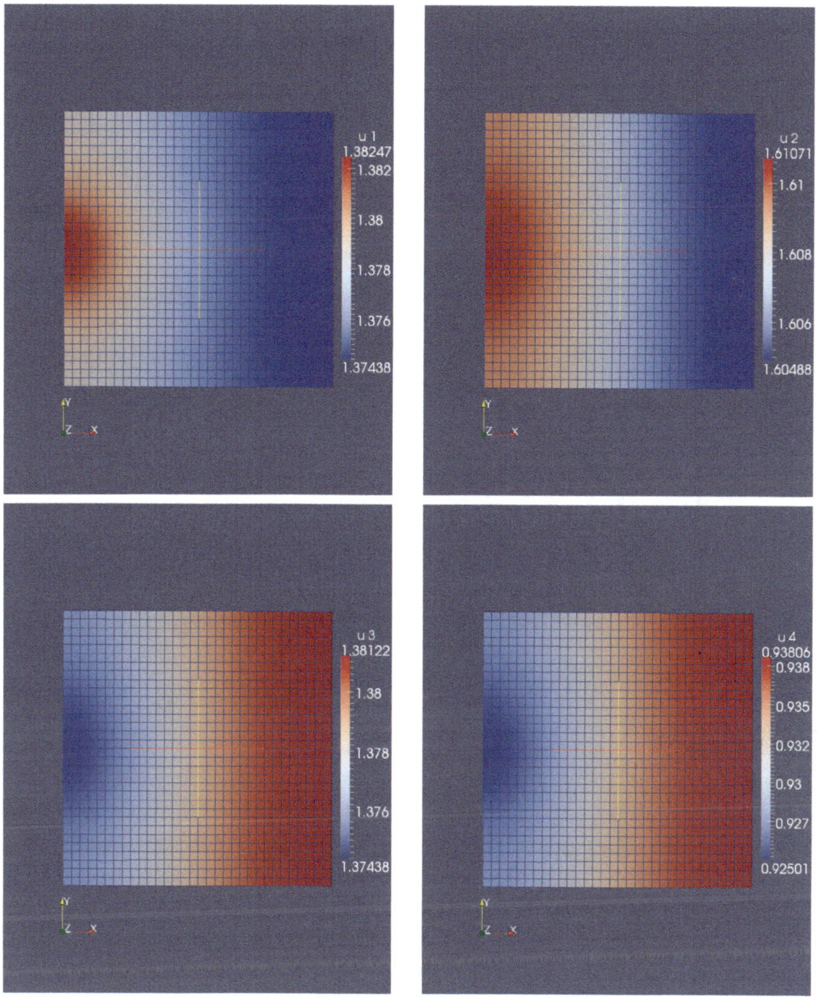

Figure 5. Clusters behavior close to the exit. The case of u_1–u_4.

In Figure 5, we see the mass escaping from the clusters u_1–u_4 in the neighborhood of the exit. Note the dramatic change in u_1 compared to what happens with the other group sizes. It is visible that large group have to stay in the queue until the small groups exit.

On the other hand, we can see in Figure 6 how the crowd breakage directly influences the outward flux. Essentially, a faster splitting of the groups tends to increase the averaged outgoing (evacuation) flux. This effect is due to our choice of boundary conditions at the exit. We mentioned

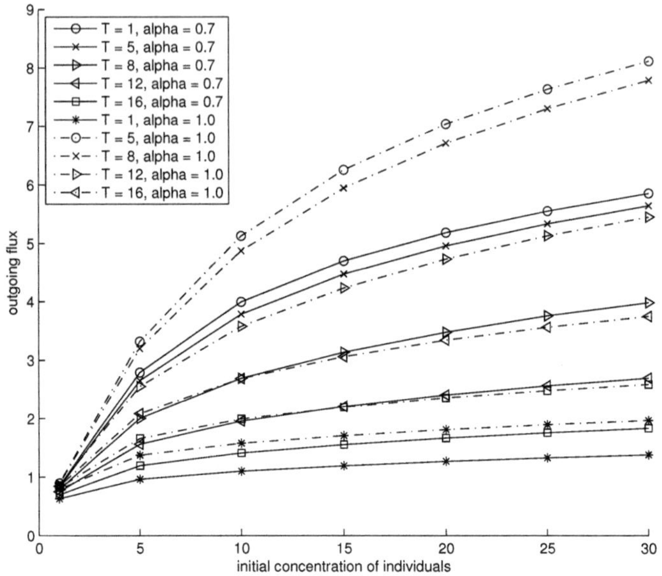

Figure 6. Comparison of outgoing flux for different values of degradation coefficients α.

in Section 2 that we expect that the way the threshold T intervenes in the definition of the aggregation coefficients β_i (compare (12)) essentially affects the maximum allowable group size. We can now see that close to the steady state situation, such situation happens. This effect is pointed out in Figure 4; the picture suggests that the mass of pedestrians piles-up in structures whose maximum lie around T.

3 A lattice model for the reverse *mosca cieca* game

3.1 Microscopic dynamics

Using the lattice model presented in this section, we explore the effects of the microscopic non–exclusion on the overall exit flux (evacuation rate). More precisely, we look again at social thresholds and study this time the effect of the *buddying threshold* (of no–exclusion per site) on the dynamics of the crowd and investigate to which extent such approach confirms the following pattern revealed by investigations on real emergencies and also emphasized in Section 2: *If the evacuees tend to cooperate and act altruistically, then their collective action tends to favor the occurrence of disasters*[5].

Question (Q1) in Paragraph 1 drives any possible attempt of modeling pedestrians motion. In this section we show how an answer to this question can be setup by using a stochastic point of view.

Our reference scenario is here a microscopic one: Imagine to be one of the individuals in a dark (possibly crowded) corridor trying to save your life by quickly reaching one of the exits. You cannot see anything and, maybe, you do not have any *a priori* knowledge of the geometry of the corridor you have to exit from. It is not difficult to imagine that you will not be able to keep a constant direction of motion and that, in any case, it will be not chosen via some neat reasoning, but you will essentially chose it at random on the basis of what other people shout and scream. In some sense your motion will closely resemble that of the blinded kid playing *mosca cieca*[6,7] with his friends.

This simple remark triggered us to propose a stochastic model for the pedestrian motion in no–visibility areas based on a random walk scheme Cirillo and Muntean (2013, 2012). The random walk rule has been introduced by taking into account a possible interaction between the individuals, see the question (Q2) in Section 1.

Pedestrians move freely inside the corridor and like to buddy with people they accidentally meet at a certain point (site). The more people are localized at a certain site, the stronger the preference to attach to it. However if

[5] Note that,due to the lack of visibility, anticipation effects (see Suma et al. (2012)) and drifts (see Guo et al. (2012)) are expected to play no role in evacuation.

[6] *Mosca cieca* means in Italian *blind fly*. It is the Italian name of a traditional children's game also known as *blind man's buff* or *blind man's bluff*. The game is played in a spacious free of dangers area in which one player, the "mosca", is blindfolded and moves around attempting to catch the other players without being able to see them. Other players try to avoid him; they make fun of the "mosca" inducing him to change direction. When one of the player is finally caught, the "mosca" has to identify him by touching is face and if the person is correctly identified he becomes the "mosca". Interestingly, the game has inspired significantly satiric literature (Manzoni, 1909; Muşatescu, 1978; Богданов, 2001). Our model tackles a reverse mosca cieca game – all the players (pedestrians) cluster around, as if they were blindfolded, trying to catch the (invisible) exit. Note that the game is actually international *жмурки* (Russian), *baba-oarba* (Romanian), *Blindekuh* (German) ...

[7] The picture in Figure 7 is taken from http://commons.wikimedia.org/wiki/File:Jongensspelen_14.jpg.

Figure 7. The blind man's buff game (the *mosca cieca* (*ital.*) game).

the number of people at a site reaches a threshold, then such site becomes not attracting for eventually new incomers.

Our lattice model provides a not so nice answer: In many situations, it seems much better not to cooperate[8]. More precisely, in Section 3.2, we will see that simulations indicate to

- cooperate with one person at time;
- cooperate with more than one person only if the number of evacuees in the corridor is not too large.

Based on this idea we have announced in Cirillo and Muntean (2012) and then presented in details in Cirillo and Muntean (2013) a model[9] for the motion of pedestrians governed by the following four mechanisms:

(A1) in the core of the corridor, people move freely without constraints;

(A2) the boundary is reflecting;

(A3) people are attracted by bunches of other people up to a threshold (*buddying mechanism*);

[8] "Cooperation" means in this setting "buddying" - the basic gregarious tendency. Our current modeling approach does not yet allow the particles to influence each other. We refer the reader to Eggels (2013) for a setting where particles do exchange mass (as a measure of "confidence") not only momentum.

[9] The model proposed in the paper is slightly more complicated, for instance there it is taken into account the possibility to tune the interaction between the pedestrians and the wall of the corridor

(A4) people are blind in the sense that there is no drift (desired velocity) leading them towards the exit.

Let $\Lambda \subset \mathbb{Z}^2$ be a finite square with odd side length L. We refer to this as the *corridor*. Each element x of Λ will be called a *cell* or *site*. The external boundary of the corridor is made of four segments made of L cells each; the point at the center of one of these four sides is called *exit*. Let N be positive integer denoting the (total) *number of individuals* inside the corridor Λ. We consider the state space $X := \{0, \dots, N\}^{\Lambda}$. For any state $n \in X$, we let $n(x)$ be the *number of individuals* at cell x.

We define a Markov chain n_t on the finite state space X with discrete time $t = 0, 1, \dots$. The parameter of the process is the integer (possibly equal to zero) $T \geq 0$ called *threshold*. We finally define the function $S : \mathbb{N} \to \mathbb{N}$ such that

$$S(k) := \begin{cases} 1 & \text{if } k > T \\ k+1 & \text{if } k \leq T \end{cases}$$

for any $k \in \mathbb{N}$. Note that for $k = 0$ we have $S(0) = 1$.

The transition matrix of the Markov chain is specified by assigning the stochastic rule according to which the individuals move on the lattice. At each time t, the N individuals move simultaneously within the corridor according to the rules that will be specified in the following. These rules depend on the location of the pedestrian, we have to distinguish among four cases: bulk, corner, neighboring the wall, and neighboring the exit (see Figure 8. In the bulk: the probability for a pedestrian at the site x to jump to one of the four neighboring sites $y_1, \dots, y_4$ is

$$\frac{S(n(y))}{S(n(x)) + S(n(y_1)) + \cdots + S(n(y_4))}.$$

In a corner: the probability for a pedestrian at the site x to jump to one of the two neighboring sites y_1 and y_2 is

$$\frac{S(n(y))}{S(n(x)) + S(n(y_1)) + S(n(y_2))}.$$

In a site close to the boundary: the probability for a pedestrian at the site x to jump to one of the three neighboring sites y_1, y_2, and y_3 is

$$\frac{S(n(y))}{S(n(x)) + S(n(y_1)) + S(n(y_2)) + S(n(y_3))}.$$

In front of the exit: the probability for a pedestrian at the site x to jump to one of the three neighboring sites y_1, y_2, and y_3 in the bulk is

$$\frac{S(n(y))}{S(n(x)) + S(n(y_1)) + S(n(y_2)) + S(n(y_3)) + (T+1)},$$

whereas the probability to exit is

$$\frac{T+1}{S(n(x))+S(n(y_1))+S(n(y_2))+S(n(y_3))+(T+1)}.$$

In all the cases described above, the probability for the individual to stay at the same site x (not to move) is $S(n(x))$ divided by the corresponding normalization denominator.

Figure 8. Schematic description of the different situation considered in the definition of the transition matrix.

The dynamics is then defined as follows: at each time t, the position of all the individuals on each cell is updated according to the probabilities defined above. If one of the individuals jumps on the exit cell a new individual is put on a cell of Λ chosen randomly with the uniform probability $1/L^2$.

3.2 Playing games on lattices

The possible choices for the parameter T correspond to two different physical situations. For $T = 0$ the function $S(k)$ is equal to one whatever the occupation numbers. This means that each individual has the same probability to jump to one of its nearest neighbors or to stay on his site. This is the independent symmetric random walk case with not zero resting probability. The second physical case is $T > 0$. For instance, $T = 1$ means mild buddying, while $T = 100$ would express an extreme buddying. No simple exclusion is included in this model: on each site one can cluster as many particles (pedestrians) as one wants. The basic role of the threshold is the following: The weight associated to the jump towards the site x increases from 1 to $1 + T$ proportionally to the occupation number $n(x)$ until $n(x) = T$, after that level it drops back to 1. Note that this rule is given on weights and not to probabilities. Therefore, if one has T particles at y and T at each of its nearest neighbors, then at the very end one will

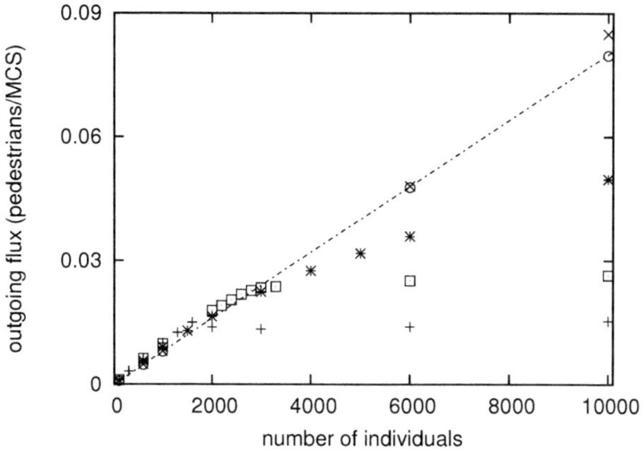

Figure 9. Averaged outgoing flux vs. number of pedestrians. The symbols $\circ$, $\times$, $*$, $\square$, and $+$ refer respectively to the cases $T = 0, 1, 5, 30, 100$. The straight line has slope 8×10^{-6} and has been obtained by fitting the Monte Carlo data corresponding to the case $T = 0$.

have that the probability to stay or to jump to any of the nearest neighbors is the same. Differences in probability are seen only if one of the five (sitting in the core) sites involved in the jump (or some of them) has an occupation number large (but smaller than the threshold).

In Cirillo and Muntean (2013), we have studied numerically this model for $T = 1, 2, 5, 30$, and 100. The Monte Carlo simulations have been all performed for $L = 101$. For each value of the threshold we have studied the cases $N = 100, 600, 1000, 6000, 10000$. For the choices $T = 30$ and $T = 100$ we have also analyzed the cases $N = 2000, 2200, 2400, 2600, 2800, 3000, 3300$ and $N = 1300, 1600, 2000, 3000$, respectively.

The main quantity of interest that one has to compute is the *average outgoing flux* that is to say the ratio between the number of individuals which exited the corridor in the time interval $[0, t_{\mathrm{f}}]$ and t_{f}. This quantity fluctuates in time, but for times large enough it approaches a constant value. In order to observe relative fluctuations smaller than 10^{-2} we had to use $t_{\mathrm{f}} = 5 \times 10^6$. To capture the extreme buddying case $T = 100$, we used $t_{\mathrm{f}} = 1.5 \times 10^7$.

Figure 9 depicts our results, where the averaged outgoing flux is given as a function of the number of individuals. At $T = 0$, that is when no buddying between the individuals is put into the model, the outgoing flux results proportional to the number of pedestrians in the corridor; indeed the data represented by the symbol $\circ$ in Figure 9 have been perfectly fitted by a straight line.

The appearance of the straight line was expected in the case $T = 0$ since in this case the dynamics reduces to that of a simple symmetric random walk with reflecting boundary conditions; see also the straight line in Figure 1 (where we suspect that, microscopically, something very similar microscopically happens). This effect was studied rigorously in the one–dimensional case and via Monte Carlo simulations in dimension two in Andreucci et al. (2011). The order of magnitude of the slope can be guessed with a simple argument Andreucci et al. (2012): the typical time needed by the walker, started at random in the lattice, to reach the site facing the exit is of order of

$$\left(\frac{1}{6}L\right)^2 \times 4L = \frac{1}{9}L^3.$$

The first term is the square of the average distance of a point inside a square of side length L from the boundary of the square itself and the second one is the number of times the walker has to visit the internal boundary before facing the exit. Hence

$$\text{outgoing flux} = \frac{1}{t_\mathrm{f}} N \frac{t_\mathrm{f}}{L^3/9} = \frac{9}{L^3} N = 8.73 \times 10^{-6}\, N.$$

When a weak buddying effect is introduced in the model, that is in the case $T = 1$, we find that if the number of individuals is small enough, say $N \leq 6000$, the behavior is similar to the one measured in the absence of buddying ($T = 0$). At $N = 10000$, on the other hand, we measure a larger flux; meaning that in the *crowded* regime small buddying favors the evacuation of the corridor [i.e. it favors the finding of the door].

The picture changes completely when buddying is increased. To this end, see the cases $T = 5, 30, 100$. The outgoing flux is slightly favored when the number of individuals is low and strongly depressed when it this becomes high. The value of N at which this behavior changes strongly depends on the threshold parameter T.

The question remains:

Why does the disaster occur at large threshold and large density?

It is not straightforward to understand how the model behaves in this regime. Inspired by theory behind particles percolation in porous media, one possible natural explanation would be that individuals cluster in bunches and that the resulting dynamics is characterized by the motion of these huge

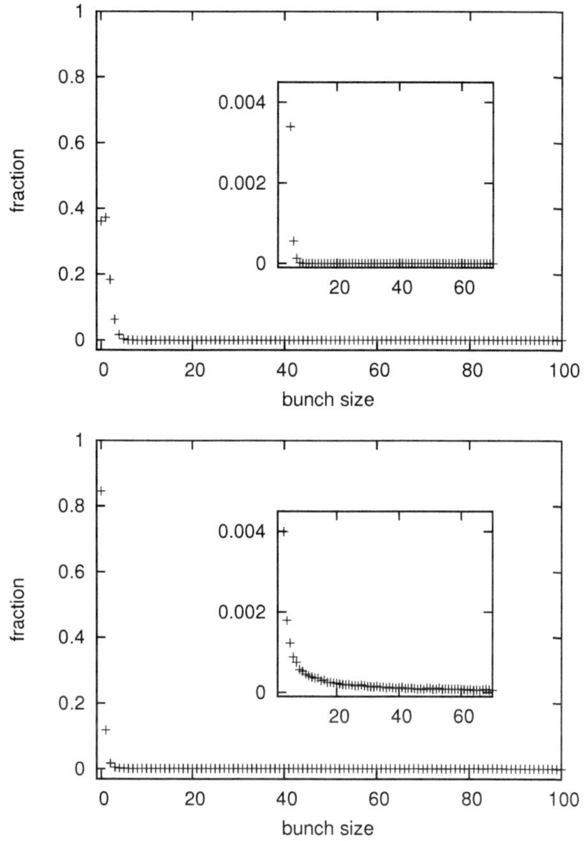

Figure 10. Histogram of the size of the bunch of people occupying the center of the lattice for $N = 10000$, $T = 0$ (left), and $T = 100$ (right).

groups. At the moment we do not know if this explanation is the right one. In order to support it at least partially, we have computed the histogram of the size of the bunch at the center of the corridor; see Figure 3.2. Here we compare the cases $T = 0$ and $T = 100$ for $N = 10000$ individuals. The histogram has been constructed by running a 10^6 long simulation. The picture does suggest that the bunch formation is negligible in the former case while in the latter it is a possible mechanism.

Now, we can summarize our conclusions based on this microscopic model. Through a novel lattice model we have examined the effect of buddying mechanisms on the efficiency of evacuation in a smoky corridor (no–visibility area). With respect to the outgoing flux measured in absence of group formation, our model predicts that

- the existence of many small groups (threshold T equal to one) favors the exit efficiency (compare points and straight line in figure 9: straight line is essentially the not–buddying case);
- strong gregariousness favors the exit efficiency only if the number of evacuees is small enough;
- the larger the threshold, the more dramatic is this effect.

In Heliavaara et al. (2012), the authors present an experiment whose purpose was to study evacuees exit selection under different behavioral objectives. The evacuation (egress) time of the whole crowd turned out to be shorter when the evacuees behave egoistically instead of behaving cooperatively. This is rather intriguing and counter intuitive fact, and it is very much in the spirit of the effect of the threshold T we observed above.

Note that for low densities the buddying mechanism increases the outgoing flux, whereas at large densities the scenario is dramatic: isolated individuals may turn to have a bigger escape chance than a large group around a leader [behavior recommended by standard manuals on evacuation strategies, see e.g. NIB (2009), p. 122.]. This suggests that evacuation strategies should not rely too much only on the presence of a leader; see Katsikopoulos and King (2010) for a related scenario.

4 Open issues

This research opens a series of fundamental questions. Some of them connect to the psychology of pedestrian groups that are essentially driven by features, behaviors, and not necessarily by desired velocities encoding the information on the location and accessibility of the exits. Some other questions are more general and refer to effect of the threshold on the general behavior of solutions to both cellular-like automata (lattice systems) as well as on Becker-Döring-like systems of differential equations (continuum systems).

We conclude the paper by enumerating a few detailed questions as well as less crystalized but promising links to other fields of science:

(i) Is there a direct link between the models (or variants on the same theme) presented in Section 2 and in Section 3? Can one derive in the many-particle limit (i.e. $N \to \infty$)) Becker-Döring-like equations having as departure point a particle system with threshold dynamics governing the interactions? We expect that a few hints can be taken over from Großkinsky et al. (2005) at least in what the moderately stochastically interacting particle limit case is concerned. Note that some ideas on how one could possibly treat simple interacting-particle systems with threshold are also anticipated in Bodineau et al. (2010), e.g., in the context of modeling batteries. For the passage from the Becker-Döring-like system to the corresponding continuity equation, ideas from Niethammer (2004) may turn to be useful.

(ii) We do not know yet how pedestrians should behave if they don't posses any information on the location of the exit. Difficult questions are: What is the right type of behavior in the dark? or How do people behave close to walls? To choose what is the best strategy for moving [e.g. cooperation (grouping, buddying, etc.) *versus* selfishness (walking away from groups)] one may also wish to explore basic aspects of the dynamics of non-momentum conserving inelastic collisions. Billiard dynamics, or biased billiards like those modeling the prisoner's dilemma, or broader contexts involving stochastic game theory (see Szilagyi (2003)), perhaps involving non-standard (strongly non-Gaussian) scenarios, where energy can be exchanged between particles in a non-standard way need to be studied Eggels (2013). Recall that the Newtonian principle of action and reaction is not necessarily true anymore in this framework; see Haret (1910).

(iii) A quite similar pile-up effect to the one seen in Figure 4 appears as a result of the motion of edge dislocations on slip planes in steel plasticity. The dislocations are repulsively interacting defects naturally arising in the crystalline structure of materials (here dual phase steels). Their motion is typically accelerated by the action of a macroscopic stress. As result of this, the dislocations are pushed towards a piling-up in the boundary later present at the interface between the strong and weak material phase; see Geers et al. (2013); van Meurs et al. (2013) for mathematical evidence on the formation of the pile-up starting off from a suitably interacting particle system. Is there a hidden threshold mechanism responsible for the formation of the pile-up of dislocations? We suspect that the high contrast between the stiffnesses of the two steel phases is the responsible threshold. We plan to use a rigorous upscaling/homogenization procedure to shed more light on connecting density thresholds (high-contrast) with pilling-ups.

(iv) To which extent cooperation is profitable? is a basic question studied recently for instance in Curşeu et al. (2013).psychologists and socio-econo- physicists. neglecting the effect of population size, thresholds and boundary conditions, The authors of Curşeu et al. (2013) are pointing out the superiority of collaborative interaction rules as compared to follow-the-leader type of interactions, making clear connections between concepts like group rationality and deliberative democracy. From yet a different perspective, this subject is intimately connected to the dynamics of opinions (cf. e.g. the work by S. Galam; to get a hint on this see Galam (2011); Martins and Galam (2012) and references cited therein) as indicated also in Moshman (March 13, 2013) (in the spirit that deliberative democracy outreasons enlightened dictatorship). One could stretch more this idea towards eventual links to percolation theory applied this time not to a porous media setting, but rather to dynamically evolving networks (societies). We refer the reader to van Santen et al. (2010), for some preliminary thoughts around the idea of percolation thresholds occurring in structured social systems.

(v) Both the lattice system and the population balances approach à la Becker-Döring share many similarities. However, there are a few essential differences between the two approaches. An important one is the following: For small N, the presence of the threshold T seems to be beneficial for the particles leaving the lattice system; however this effect is lost completely in the Becker-Döring approach (compare Figure 5). This seems to be due to the choice of boundary conditions in the continuum system. On the other hand, we conjecture that the continuum limit of the lattice system is a sort of non-linear diffusion equation with inherited threshold, while we see that the Becker-Döring system is not emphasizing the threshold effects when changing the size (or nonlinearity) of the effective diffusion coefficient (see e.g. Figure 3). The challenging question is here: Derive (and then prove rigorously) the mean-field limit for the lattice system. Alternatively, one can reformulate the lattice model in terms of myopic random walkers in an exclusion process in the spirit of Landman and Fernando (2011) and then prove rigorously the validity of the corresponding mean-field model (a porous media-like equation).

(vi) Based on our working experience with continuum models with distributed microstructures, we expect that it is possible to couple the two models for groups dynamic within a single multiscale framework. The challenge here is to establish the right micro-macro transmission condition (in this case, a discrete-to-continuum coupling). We believe that steps in this direction are possible, inspired for instance by the way the human language is treated in Mitchener (2010) as a hybrid system.

Acknowledgments

We thank Anne Eggels, Joep Evers, Francesca Nardi and Rutger van Santen (Eindhoven), Petre Curşeu (Tilburg) as well as Errico Presutti (Rome) for fruitful discussions on this and closely related topics. A.M. thanks the NWO's Complexity program (project "Correlating fluctuations across the scales"), O.K. acknowledges financial support from the EU ITN FIRST project.

A Becker-Döring system in the context of a two-scale modeling approach

A.1 Background

This section contains a brief derivation of a structured-population model, which is a special case of a multi-feature continuity equation cf. Böhm (2012). It provides a general framework for some of the equations we are dealing with. For a related derivation using densities, see Perthame (2007), e.g. At a more general scale the following considerations yield some sort of a transport equation or continuity equation, respectively, with two features being involved in the transport (also: cf. Smoluchowski (1917); Diekmann et al. (1998) et al.). In the present situation, the "location in the corridor" and the "group size" constitute the two "features". The first is a continuous, the second a discrete variable. Our aim is to derive a population-balance equation, (21), able to describe the evolution of pedestrian groups in obscured regions.

Fix $N \in \mathbb{N}$, let $\Omega \subset \mathbb{R}^2$ be the dark corridor (open, bounded with Lipschitz boundary), S - the observation time interval and $K_d := \{0, 1, 2, 3, .., N\}$ - the collection of all admissible group sizes. We say that a Y belongs to $K' \subseteq K$, if it belongs is part of some group with a size K. Furthermore, $\mathfrak{A}_\Omega := \mathfrak{B}^2(\Omega)$, $\mathfrak{A}_S := \mathfrak{B}^1(S)$ are the corresponding Borel σ-algebras with the corresponding Lebesgue-Borel measures $\lambda_x := \lambda^2$ and $\lambda_t := \lambda^1$, respectively; $\mathfrak{A}_{K_d} := \mathfrak{p}(K_d)$ is equipped with the counting measure $\lambda'_c(K) := |K'|$. We call $\lambda_{tx} := \lambda_t \otimes \lambda_x$ the *space-time measure* and set $\lambda_{txc} := \lambda_t \otimes \lambda_x \otimes \lambda_{c_j}$, $\mathfrak{A}_{\Omega K_d} := \mathfrak{A}_\Omega \otimes \mathfrak{A}_{K_d}$, $\mathfrak{A}_{S\Omega K_d} := \mathfrak{A}_S \otimes \mathfrak{A}_\Omega \otimes \mathfrak{A}_{K_d}$.

A.2 Derivation of the model

Fix $t \in S$, let $\Omega' \in \mathfrak{A}_\Omega$, $K' \in \mathfrak{A}_{K_d}$, $S' \in \mathfrak{A}_S$, introduce

$$\begin{aligned} \mu_Y(t, \Omega' \times K') := {} & \text{number of } Y's \text{ present in } \Omega' \\ & \text{at time } t \text{ and belonging to the group } K' \end{aligned} \tag{13}$$

and two *production quantities*

$$\mu_{PY\pm}(S' \times \Omega' \times K') := \text{ number of } Y's \text{ which are added} \\ \text{to (subtracted from) } \Omega' \times K' \text{ during } S' \text{ and} \\ \mu_{PY} = \mu_{PY+} - \mu_{PY-}. \tag{14}$$

Note that these numbers might be non-integer.

Given the nature of the problems we are dealing with, we postulate - as a part of the modeling-

(P1) For all $K' \in \mathfrak{A}_{K_d}, \Omega' \in \mathfrak{A}_\Omega$ and $t \in S : \mu_Y(t, \cdot \times K')$ and $\mu_Y(t, \Omega' \times \cdot)$ are measures on their respective σ-algebras $\mathfrak{A}_\Omega$ and $\mathfrak{A}_{K_d}$, respectively.

(P2) $\mu_{PY\pm}(S' \times \Omega' \times \cdot)$, $\mu_{PY\pm}(S' \times \cdot \times K')$ and $\mu_{PY\pm}(\cdot \times \Omega' \times K')$ are measures on their respective σ-algebras.

Now, we are in the position to formulate a

$$\text{Balance principle:} \\ \mu_Y(t+h, \Omega' \times K') - \mu_Y(t, \Omega' \times K') = \mu_{PY}(S' \times \Omega' \times K') \\ \text{for all } t, t+h \in S,\ \Omega' \times K' \in \mathfrak{A}_\Omega \times \mathfrak{A}_{K_d},\ S' := (t, t+h]. \tag{15}$$

Addition to $\Omega' \times K'$, modeled by μ_{PY+}, can happen by addition *inside* of $\Omega' \times K'$ as well as by fluxes *into* $\Omega' \times K'$. A similar remark applies to subtraction and μ_{PY-}. This gives rise to assume μ_{PY+} to be the sum of an interior production part, μ_{PY+}^{int}, and a flux part, μ_{PY+}^{flux}. We proceed similarly with μ_{PY-} and have, with the

$$\text{net productions } \mu_{PY}^{int} := \mu_{PY+}^{int} - \mu_{PY-}^{int} \text{ and } \mu_{PY}^{flux} := \mu_{PY+}^{flux} - \mu_{PY-}^{flux} :$$

$$\mu_{PY} = \mu_{PY}^{int} + \mu_{PY}^{flux} = \left(\mu_{PY+}^{int} - \mu_{PY-}^{int}\right) + \left(\mu_{PY+}^{flux} - \mu_{PY-}^{flux}\right). \tag{16}$$

We extend $\mu_Y(t, \cdot \times \cdot)$ and $\mu_{PY\pm}(\cdot \times \cdot \times \cdot)$ by the usual procedure to measures $\overline{\mu}_Y = \mu_Y(t, \cdot)$ and $\overline{\mu}_{PY\pm} = \overline{\mu}_{PY\pm}(\cdot)$ on the product algebras $\mathfrak{A}_\Omega \otimes \mathfrak{A}_{K_d}$ and $\mathfrak{A}_S \otimes \mathfrak{A}_\Omega \otimes \mathfrak{A}_{K_d}$, respectively.

Note that the quantities in (P1) and (P2) and the extensions are finite.

The following postulate prevents accumulation on sets of measure zero. It reads as

(P3) $\overline{\mu}_Y(t, \cdot) \ll \lambda_{xc}$ (absolutely continuous).

Therefore, for all $t \in S$ there are integrable Radon-Nikodym densities $u(t, \cdot) = \frac{d\overline{\mu}_Y(t, \cdot)}{d\lambda_{xc}}$[10], i.e.

$$\overline{\mu}_Y(t, Q') = \int_{Q'} u(t, (x, i)) d\lambda_{xc} \text{ for all } Q' \in \mathfrak{A}_{\Omega K}. \tag{17}$$

[10] Note with respect to Section 2: $u_i(t, x)$ from Section 2 corresponds to $u(t, x, i)$ here.

The absolute-continuity assumption

(P4) $\overline{\mu}_{PY}^{int} \ll \lambda_{txc}$

excludes the presence of $Y's$ on sets of λ_{txc_1}-measure zero. Moreover it assures the existence of the Radon-Nikodym density

$$f_{PY}^{int} := \frac{d\overline{\mu}_{PY}}{d\lambda_{txc}} \in L_{loc}^1(S \times \Omega \times K_d, \mathfrak{A}_{S\Omega K_d}, \lambda_{txc}). \tag{18}$$

In order to get a reasonable idea for a representation of the flux measure we consider the special case $Q' = \Omega' \times K'$ with, say, $K' = \{a, a+1, ..., b\} \in \mathfrak{p}(K_d)$. The "surface"

$$\mathfrak{F} := \Omega' \times \{a\} \cup \Omega' \times \{b\} \cup \partial\Omega' \times K'$$

is the location of any interaction with the outside of Q'. There are two locations on $\mathfrak{F}$ to enter or leave Q' from the outside - one via $\mathfrak{F}_1 := \Omega' \times \{a\} \cup \Omega' \times \{b\}$, the other one through $\mathfrak{F}_2 := \partial\Omega' \times K'$ (see Figure 11).

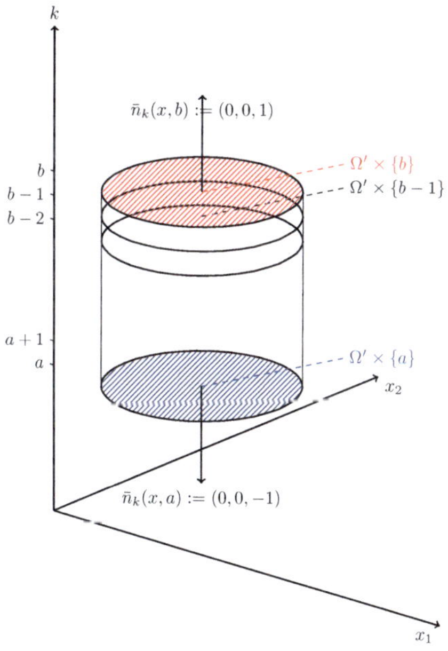

Figure 11. Special interactions regions on the surface $\mathfrak{F}$.

The unit-outward normal field $\mathbf{n} = \mathbf{n}(x, \kappa)$ on $\mathfrak{F}$ can be split into two orthogonal components, $\mathbf{n} = \mathbf{n}_x + \mathbf{n}_\kappa$, $\mathbf{n}_x = (n_x, 0)$, $\mathbf{n}_\kappa = (0, n_\kappa)$, respectively. It is $n_\kappa(x, a) = -1$, $n_\kappa(x, b) = +1$ and $n_x = n_x(x, \kappa)$ is the a.e. existing outward normal on $\partial\Omega$. Borrowing from the theory of Cauchy interactions, cf. Schuricht (2007), e.g..

(P5) we assume for all $t \in S$ the existence of two vector fields

$$\begin{aligned} j_x(t, \cdot) &: \Omega \times K_d \to \mathbb{R}^2 \\ j_\kappa &: \Omega \times K_d \to \mathbb{R} \end{aligned}$$

with

$$\overline{\mu}_{PY}^{flux} := \overline{\mu}_{PYx}^{flux} + \overline{\mu}_{PY\kappa}^{flux},$$

where

$$\begin{aligned} \overline{\mu}_{PYx}^{flux}(S' \times \Omega' \times K') &:= \int_{S'} \int_{\mathfrak{F}_2} -j_x(\tau, x, i) \cdot n_x(x, i) d\sigma_x d\lambda_c d\tau \\ &\text{and} \\ \overline{\mu}_{PY\kappa}^{flux}(S' \times \Omega' \times K') &:= \int_{S'} \int_{\Omega'} -j_\kappa(\tau, x, b) n_\kappa(x, b) \\ &-j_\kappa(\tau, x, a) n_\kappa(x, a) dx d\tau. \end{aligned} \tag{19}$$

In (19), σ_x - is the $1D$-surface (= curve length-) measure. $\overline{\mu}_{PYx}^{flux}(S' \times \Omega' \times K')$ calculates the net gain/loss of the $Y's$ in Ω' belonging to one of the size groups from K' due to physical motion from/to the outside of Ω' into/out of Ω'.

Furthermore, $\overline{\mu}_{PY\kappa}^{flux}(S' \times \Omega' \times \{i\})$ calculates the net gain/loss of the $Y's$ in Ω' belonging to the size group labelled by i due to reasons within K. Since, in the given situation of Section 2, there is no interaction with groups of size $\kappa > N$ or $\kappa < 0$ (these group sizes are not admissible!), we have to require

$$j_\kappa(t, x, 0) = j_\kappa(t, x, N) = 0 \quad \text{for all } t \in S,\ x \in \Omega. \tag{20}$$

Introducing the *discrete partial derivative* by

$$\partial_i^d j_\kappa(t, x, i) := j_\kappa(t, x, i+1) - j_\kappa(t, x, i),\ i \in K$$

and assuming u, f_{PY}^{int}, $\operatorname{div}_x j_x$ and $\partial_\kappa^d j_\kappa$ to be sufficiently regular, we obtain

$$\begin{aligned} \overline{\mu}_{PY}^{flux}(S' \times Q') = &\int_{S'} \int_{Q'} -\operatorname{div}_x(j_x(\tau, x, i) d\lambda_c dx d\tau \\ &+ \int_{S'} \int_{Q'} -\partial_i^d j_\kappa(t, x, i) d\tau dx d\lambda_c. \end{aligned}$$

Combining (15) - (19), Fubini's theorem, and division by h, imply

$$\begin{aligned}&\int_{Q'} \frac{1}{h}\left(u(t+h,x,i)-u(t,x,i)\right)dxd\kappa\\ =&\int_{Q'} \frac{1}{h}\int_t^{t+h}\left(f_{PY}^{int}(\tau,x,i)-\left(\operatorname{div}_x j_x(\tau,x,i)+\partial_i^d j_\kappa(t,x,i)\right)\right)d\tau dx d\lambda_c.\end{aligned}$$

Under appropriate smoothness conditions on u, f_{PY}^{int}, j_x and j_κ we obtain in the limit $h\to 0$ (the classical continuity equation with a slightly different interpretation of the entries)

$$\frac{\partial u}{\partial t}(t,x,i)+\left(\operatorname{div}_x j_x+\partial_i^d j_\kappa(t,x,i)\right)=f_{PY}^{int}(t,x,i). \tag{21}$$

A.3 Connection with the model in Section 2:

In order to obtain a workable model, one has to specify the flux vectors j_x and j_κ as well as f_{PY}^{int}. In Section 2 this has been done in (1) to (6) by setting

$i=1,...,N$ (there) $=$ $i=1,...,N$ (here), $u_i(t,x)$ (there) $=$ $u(t,x,i)$ (here), $-D_i\nabla_x u_i(t,x)$ (there) $=$ $j_x(t,x,i)$ (here),

$$f_{PY}^{int}(t,x,i)(\text{here})=\begin{cases}\sum_{i=1}^N \alpha_i u_i-\sum_{i=1}^N +\beta_i u_i u_1 & \text{if } i=1,\\ \beta_{i-1}u_{i-1}u_1-\beta_i u_i u_1 & \text{if } i\in\{2,...,N-1\},\\ \beta_N u_{N-1}u_1 & \text{if } i=N,\end{cases}$$

respectively.

The discrete derivative $j_\kappa(t,x,i)$ (here) corresponds to

$$\begin{aligned}j_\kappa(t,x,i)&=-\alpha_i u_i(t,x),\ i=1,2,...,N-1,\\ j_\kappa(t,x,0)&=j_\kappa(t,x,N)=0.\end{aligned}$$

A.4 Derivation of the model in Section 2:

Specifying $j_x(t,x,i)$ as some sort of a diffusion flux in the manner above means: Individual groups of size i recognize whether a group of the same size is in their immediate neighborhood and they tend to avoid moving into the direction of such groups. Employing a Fickian law seems to be the simplest way to model this. f_{PY}^{int} models interactions ($=$ merging) between groups of size $i\in K$ and "groups" of size $i=1$: If a single (i.e. a group of size one) hits a group of size $i<N$, then it might happen, that this single merges with the group. This turns the group into a group of size $i+1$ and leads to a "gain" for groups of size $i+1$ (modeled by $+\beta_i u_i u_1$) and a loss for groups of size i (modeled by $-\beta_i u_i u_1$). In any such joining situation the group with $i=1$ looses members (modeled by $-\sum_{i=2}^N \beta_i u_i u_1$). Note, that

this model allows only for direct interaction between groups of size i with groups of size 1! The α-terms model some "degradation" effect: It might happen, that an individual leaves a group of size $i \geq 2$. This leads to a loss for the groups of size i (modeled by $-\alpha_i u_i$), a gain for the groups of size $i-1$ and a gain for the groups with $i = 1$ (modeled by $\sum_{i=2}^{N} \alpha_i u_i$). α_i, $\beta_i \geq 0$ and $D_i > 0$ are empirical and assumed to be constant.

Note: In the abstract approach the degradation terms express a flux rather than a volume source or sink. In the same way as aging can be seen as a flux ("people change their age group by aging with (speed 1)") Y's change their size group by "degradation" of their group. Nevertheless: For fixed i, the expressions $\alpha_i u_i$ and $\alpha_{i-1} u_{i-1}$ still remain "volume sources" and "sinks", respectively. It's just two different ways to look at the same thing.

Bibliography

Basisopleiding Bedrijfshulpverlener. NIBHV – Nederlands Instituut voor Bedrijfshulpverlening, Rotterdam, 2009.

D. Andreucci, D. Bellaveglia, E. N. M. Cirillo, and S. Marconi. Monte Carlo study of gating and selection in potassium channels. *Phys. Rev. E*, 84: 021920, 2011.

D. Andreucci, D. Bellaveglia, E. N. M. Cirillo, and S. Marconi. Effect of intracellular diffusion on current–voltage curves in potassium channels. *arXiv: 1206.3148*, 2012.

T. Bodineau, B. Derrida, and J. L. Lebowitz. A diffusive system driven by a battery or by a smoothly varying field. *Journal of Statistical Physics*, 140(4):648–675, 2010.

M. Böhm. *Lecture Notes in Mathematical Modeling.* Universität Bremen, Germany, 2012. Fachbereich Mathematik und Informatik,.

O. Chepizhko, E. Altmann, and F. Peruani. Collective motion in heterogeneous media. *Phys. Rev. Lett.*, 2013:to appear, 2013.

E. N. M. Cirillo and A. Muntean. Dynamics of pedestrians in regions with no visibility – a lattice model without exclusion. *Physica A: Statistical Mechanics and its Applications*, 392(17):3578 – 3588, 2013.

E. N. M. Cirillo and A. Muntean. Can cooperation slow down emergency evacuations? *Comptes Rendus Mecanique*, 340:626–628, 2012.

P. L. Curşeu. Group dynamics and effectiveness: A primer. In S. Boros, editor, *Organizational Dynamics*, chapter 7, pages 225–246. Sage, London, 2009.

P. L. Curşeu, R. J. Jansen, and M. M. H. Chappin. Decision rules and group rationality: Cognitive gain or standstill? *PLoS ONE*, 8(2):e56454, 2013.

O. Diekmann, M. Gyllenberg, J. A. J. Metz, and H. R. Thieme. On the formulation and analysis of general deterministic structured population models I. Linear theory. *Journal of Mathematical Biology*, 36(4):349–388, 1998.

J. R. G. Dyer, A. Johansson, D. Helbing, I. D. Couzin, and J. Krause. Leadership, consensus decision making and collective behaviour in humans. *Philosophical Transactions of the Royal Society: Biological Sciences*, 364: 781–789, 2009.

J. T. Edward. Molecular volumes and the Stokes-Einstein equation. *Journal of Chemical Education*, 47(4):261, 1970.

A. Eggels. Social billiards – a novel approach to crowd dynamics in dark (Bachelor thesis, TU Eindhoven), 2013.

J. H. M. Evers and A. Muntean. Modeling micro-macro pedestrian counterflow in heterogeneous domains. *Nonlinear Phenomena in Complex Systems*, 14(1):27–37, 2011.

Z.-M. Fang, W.-G. Song, J. Zhang, and H. Wu. A multi-grid model for evacuation coupling with the effects of fire products. *Fire Technology*, 48:91–104, 2012.

S. Galam. Collective beliefs versus individual inflexibility: The unavoidable biases of a public debate. *Physica A-Statistical Mechanics and Its Applications*, 390:3036–3054, 2011.

M.G.D. Geers, R.H.J. Peerlings, M.A. Peletier, and L. Scardia. Asymptotic behaviour of a pile-up of infinite walls of edge dislocations. *Arch. Ration. Mech. Analysis*, 209(2):495–539, 2013.

S. Großkinsky, C. Klingenberg, and K. Ölschläger. A rigorous derivation of Smoluchowski's equation in the moderate limit. *Stochastic Analysis and Applications*, 22(1):113–141, 2005.

M. Z. Guo, G. C. Papanicolaou, and S. R. S. Varadhan. Nonlinear diffusion limit for a system with nearest neighbor interactions. *Comm. Math. Phys.*, 118(1):31–59, 1988.

X. Guo, J. Chen, Y. Zheng, and J. Wei. A heterogeneous lattice gas model for simulating pedestrian evacuation. *Physica A: Statistical Mechanics and its Applications*, 391(3):582 – 592, 2012.

S. Haret. *Mécanique sociale.* Gauthier-Villars, Paris, 1910.

D. Helbing and P. Molnar. Social force model for pedestrian dynamics. *Physical Review E*, 51(5):4282–4286, 1995.

S. Heliavaara, J.-I. Kuusinen, T. Rinne, T. Korhonen, and H. Ehtamo. Pedestrian behavior and exit selection in evacuation of a corridor: An experimental study. *Safety Science*, 50(2):221 – 227, 2012.

T. Jin. Visibility through fire smoke. *Journal of Fire and Flammability*, 9: 135–155, 1978.

T. Jin and T. Yamada. Irritating effects of fire on visibility. *Fire Science and Technology*, 5:79–90, 1985.

K. V. Katsikopoulos and A. J. King. Swarm intelligence in animal groups: When can a collective out-perform an expert? *PLOS One*, 5:11, 2010.

A. Kirchner and A. Schadschneider. Simulation of evacuation processes using a bionics-inspired cellular automaton model for pedestrian dynamics. *Physica A: Statistical Mechanics and its Applications*, 312(1-2):260 – 276, 2002.

M. Kobes, I. Helsloot, B. de Vries, and J. G. Post. Building safety and human behaviour in fire: A literature review. *Fire Safety Journal*, 45(1): 1 – 11, 2010.

O. Krehel, A. Muntean, and P. Knabner. On modeling and simulation of flocculation in porous media. In A.J. Valochi (Ed.), Proceedings of XIX International Conference on Water Resources. (pp. 1-8) CMWR, University of Illinois at Urbana-Champaign, 2012.

K. A. Landman and A. E. Fernando. Myopic random walkers and exclusion processes: Single and multispecies. *Physica A: Statistical Mechanics and its Applications*, 390:3742 – 3753, 2011.

G. Le Bon. *La psychologie des foules*. The Echo Library, Middlesex, 2008.

A. Manzoni. *I Promessi Sposi (The betrothed)*. P. F. Collier and Son Comp., New York, 1909.

A. C. R. Martins and S. Galam. The building up of individual inflexibility in opinion dynamics. *CoRR*, abs/1208.3290, 2012.

W. Mitchener. Mean-field and measure-valued differential equation models for language variation and change in a spatially distributed population. *SIAM J. Math. Anal.*, 42(5):1899–1933, 2010.

D. Moshman. Deliberative democracy outreasons enlightened dictatorship. Politics section, Huffington post, March 13, 2013.

V. Muşatescu. *Extravagantul Conan Doi: De-a v-aţi ascunselea. De-a baba oarba*. Cartea Românească, 1978.

B. Niethammer. Macroscopic limits of the Becker-Doring equations. *Commun. Math. Sci.*, 2(1):85 – 92, 2004.

B. Perthame. *Transport Equations in Biology*. Frontiers in Mathematics. Birkhauser, Basel, Boston, Berlin, 2007.

B. Piccoli and A. Tosin. Time-evolving measures and macroscopic modeling of pedestrian flow. *Arch. Ration. Mech. Anal.*, 199(3):707–738, 2011.

A. Portuondo y Barceló. *Apuntes sobre Mecánica Social*. Establecimiento Topográfico Editorial, Madrid, 1912.

E. Presutti. Personal communication, 2013.

N. Ray, A. Muntean, and P. Knabner. Rigorous homogenization of a Stokes-Nernst-Planck-Poisson system. *Journal of Mathematical Analysis and Applications*, 390(1):374 – 393, 2012.

A. Schadschneider, W. Klingsch, H. Kluepfel, T. Kretz, C. Rogsch, and A. Seyfried. Evacuation dynamics: Empirical results, modeling and applications. In R. A. Meyers, editor, *Encyclopedia of Complexity and System Science*, volume 3, pages 31–42. Springer Verlag, Berlin, 2009.

A. Schadschneider, D. Chowdhury, and K. Nishinari. *Stochastic Transport in Complex Systems*. Elsevier, 2011.

F. Schuricht. A new mathematical foundation for contact interactions in continuum physics. *Arch. Ration. Mech. Anal.*, 1984:169–196, 2007.

M. Smoluchowski. Versuch einer mathematischen Theorie der Koagulationskinetik kolloider Lösungen. *Z. Phys. Chem.*, 92:129–168, 1917.

Y. Suma, D. Yanagisawa, and K. Nishinari. Anticipation effect in pedestrian dynamics: Modeling and experiments. *Physica A: Statistical Mechanics and its Applications*, 391:248 – 263, 2012.

M. N. Szilagyi. Simulation of multi-agent Prisoners' Dilemmas. *Syst. Anal. Model. Simul.*, 43(6):829–846, June 2003.

C. M. Topaz and A. L. Bertozzi. Swarming patterns in a two-dimensional kinematic model for biological groups. *SIAM J. Appl. Math.*, 65:152–174, 2004.

P. van Meurs, A. Muntean, and M. A. Peletier. Upscaling of dislocations walls in finite domains. unpublished, Eindhoven University of Technology, 2013.

R. van Santen, D. Khoe, and B. Vermeer. *2030: Technology That Will Change the World*. Oxford University Press, UK, 2010.

Y. Zheng, B. Jia, X.-G. Li, and N. Zhu. Evacuation dynamics with fire spreading based on cellular automaton. *Physica A: Statistical Mechanics and its Applications*, 390:3147–3156, 2011.

К. Богданов. *Игра в жмурки: сюжет, контекст, метафора*. Искусство-СПБ, 2001.

Stochastic competition between two populations in space

Simone Pigolotti[1], Roberto Benzi[2], Mogens H. Jensen[3], Prasad Perlekar[4], and Federico Toschi[5]

[1] Dept. de Fisica i Eng. Nuclear, Universitat Politecnica de Catalunya Edif. GAIA, Rambla Sant Nebridi s/n, 08222 Terrassa, Barcelona, Spain.

[2] Dipartimento di Fisica, Universita' di Roma "Tor Vergata" and INFN, via della Ricerca Scientifica 1, 00133 Roma, Italy.

[3] The Niels Bohr Institut, University of Copenhagen, Blegdamsvej 17, DD-2100 Copenhagen, Denmark.

[4] Tata Institute of Fundamental Research, Centre for Interdisciplinary Sciences, 21 Brundavan Colony, Narsingi, Hyderabad 500075, India

[5] Department of Physics, Department of Mathematics and Computer Science, and J.M. Burgerscentrum, Eindhoven University of Technology, 5600 MB Eindhoven, The Netherlands and CNR-IAC, Via dei Taurini 19, 00185 Rome, Italy.

Abstract We present a model describing spatial competition between two biological populations. Individuals belonging to the two populations diffuse in space, reproduce, and die as effect of competitions; all these processes are implemented stochastically. We focus on how the macroscopic equations for the densities of the two species can be derived within the formalism of the chemical master equations. We also compare the case in which the total density of individuals is kept fixed by constraint with a case in which it can fluctuate.

1 Introduction

Competition between biological populations can be mathematically described at different levels of complexity. For example, when spatial degrees of freedom and number fluctuations are neglected, competition models are relatively easy to analyze with tools of dynamical system theory. A paradigmatic example of this case are Lotka-Volterra models, see e.g. (Murray, 2007).

However, in many biological situations, the spatial distribution of the populations cannot be neglected, so that one is forced to consider spatially explicit models. Moreover, stochasticity can be also important. This is especially the case in neutral or near-neutral conditions, where the parameters characterizing the species are the same (or nearly), and the outcome of competition is determined by chance rather than by fitness differences.

In this chapter, we analyze two spatial competition models. The first is the stepping stone model, originally introduced by Kimura (Kimura, 1953; Kimura and Weiss, 1964). A key assumption of the stepping stone model is that the sum of the number of individuals belonging to the two species is kept constant at each point in space; this assumption is relaxed in the second model (Pigolotti et al., 2012, 2013). For both models, we show how one can generally derive the macroscopic dynamic equations describing the concentrations of the two species using the formalism of the chemical master equation (see e.g. Gardiner (2004), chapter 8), which can be thought of as a generalization of the Kramers-Moyal expansion for spatially extended systems. After presenting this derivation and discussing its limits of validity for the two models, we show some analytical and numerical results in the case in which the two species are neutral, i.e. characterized by the same rates.

2 The Stepping Stone Model

The stepping stone model (Kimura, 1953; Kimura and Weiss, 1964) is a paradigmatic model for spatial population genetics. Let us consider a system made up of different islands (or "demes"), each hosting two populations, A and B. The total population of each island is a fixed parameter N_l. We denote with n the population of species A, so that the population of species B is $N_l - n$. The two populations undergo a Moran process: at a given rate, an individual is chosen at random, killed and replaced with a copy of one of the other individuals on the island, also chosen at random. To model the possibility of a selective advantage, individuals of space A are copied with a rate $\mu(1+s)$, while individuals of species B are copied at a rate μ. The parameter μ can be interpreted as an inverse generation time, while s represents the relative selective advantage of species A. The rates at which population A increases or decreases are then given by:

$$\begin{aligned} W^+(n_i) &= (1+s)\mu\frac{N_l - n_i}{N_l}\frac{n_i}{N_l} \\ W^-(n_i) &= \mu\frac{n_i}{N_l}\frac{N_l - n_i}{N_l}. \end{aligned} \tag{1}$$

For simplicity, we first discuss the well-mixed version of the model, i.e.

the dynamics on a single island, which reduces to the well-known Moran model. In the second part of this section, we will describe the one dimensional case of a linear array made up of many islands, where the rates above will be complemented by immigration/emigration rates between neighboring islands.

The definition of the rates in (1) directly leads to the following master equation

$$\frac{d}{dt}P(n_i,t) = W^+(n_{i-1})P(n_{i-1},t) + \\ +W^+(n_{i+1})P(n_{i+1},t) - [W^+(n_i) + W^-(n_i)]P(n_i,t). \quad (2)$$

The next step consists in approximating the birth-death process defined above into a Langevin equation by means of a Kramers-Moyal expansion. Formally, the master equation (2) can be written in an integral form as

$$\frac{d}{dt}P(n,t) = \int d(\Delta n)[\omega(\Delta n, n - \Delta n)P(n - \Delta n) - \omega(\Delta n, n)P(n)] \quad (3)$$

where the jump rates have been incorporated into a jump distribution function ω:

$$\omega(\delta n, n) = \delta(\Delta n - 1)W^+(n) + \delta(\Delta n + 1)W^-(n). \quad (4)$$

The trick is now to perform a Taylor expansion of Eq. (3) around $\Delta n = 0$, leading to

$$\frac{d}{dt}P(n,t) = \sum_{j=1}^{\infty} \frac{(-1)^j}{j!} \frac{d^j}{dn^j}[\alpha_j(n)P(n,t)] \quad (5)$$

where the α_j's are the moments of the jump distribution,

$$\alpha_j(n) = \int d(\Delta n)(\Delta n)^j \omega(\Delta n, n). \quad (6)$$

Assuming $N_l \gg 1$, we can introduce the new variable $f = n/N_l$. The quantity f can be interpreted as the fraction of one species: $f = 1$ means an island exclusively populated with one allele and $f = 0$ means exclusive occupation by the alternative genotype. The jumps in terms of the new variable $\delta f = \pm 1/N_l$ are now small, so that we can truncate the above expansion up to the second derivative. This yields a Fokker-Planck equation:

$$\partial_t P(f,t) = -\partial_f[\mu s f(1-f)P(f,t)] + \partial_f^2 \left[\frac{\mu f(1-f)}{N_l} P(f,t)\right] \quad (7)$$

where we neglected terms of order s/N by assuming $N \gg 1$ and $s \gg 1$. The corresponding Langevin equation is

$$\partial_t f(t) = \mu s f(1-f) + \sqrt{\frac{2\mu f(1-f)}{N_l}}\xi(t) \tag{8}$$

where $\xi(\mathbf{x}, t)$ is a Gaussian stochastic process, delta-correlated in time, $\langle \xi(t)\xi(t')\rangle = \delta(t-t')$. The nonlinearity multiplying the noise requires an interpretation in terms of the Ito calculus; this will also be the case for all generalizations we will consider in the following.

We now move to the one-dimensional case. We consider an infinite linear array of islands (or "demes"), where two neighboring islands are separated by a distance a. Each island host a total population N_l of individuals belonging to the two species A and B. Numbering the islands with an index i, we denote with n_i the population of species A in the island i, so that the population of species B will be $N_l - n_i$. The local dynamics on each island is the same as before; the only additional ingredient is that we allow neighboring island for exchanging individuals. It is convenient to call the exchange rate from an island to a neighboring one as DN/a^2, where D is an additional free parameter. We can proceed as before by performing a Kramers-Moyal expansion in each island and introducing the local fractions $f_i = n_i/N_l$. The result is a set of Langevin equations:

$$\partial_t f_i(\mathbf{x}, t) = \frac{D}{2a^2}(f_{i-1} + f_{i+1} - 2f_i) + \mu s f_i(1-f_i) + \sqrt{\frac{2\mu f_i(1-f_i)}{N_l}}\xi_i(t) \tag{9}$$

where noise sources corresponding to different islands are uncorrelated. It is now possible to (formally) take the continuum limit $a \to 0$, leading to

$$\partial_t f(\mathbf{x}, t) = D\nabla^2 f(\mathbf{x}, t) + \mu s f(1-f) + \sqrt{\frac{2\mu f(1-f)}{N}}\xi(\mathbf{x}, t) \tag{10}$$

where $N = N_l/a$: it is convenient to distinguish between N_l (the population inside a single discrete deme of the SSM) and N (the corresponding total density of individuals). Notice that N_l is a non-dimensional quantity, while N is a density, carrying units of an inverse length. In the above equation, $\xi(\mathbf{x}, t)$ is a Gaussian stochastic process, delta correlated in space and time, $\langle \xi(\mathbf{x}, t)\xi(\mathbf{x}', t')\rangle = \delta(\mathbf{x} - \mathbf{x}')\delta(t-t')$.

We conclude with a few remarks about the validity of this continuous limit. Equations (7) and (8) have been derived by means of the Kramers-Moyal equation, which strictly speaking is not a systematic expansion in a small parameters. However, the same equation can be more rigorously derived in the framework of Van Kampen's system size expansion, where the expansion parameter is $1/N_l$ (see e.g. Risken (1989); Gardiner (2004) for a discussion of this problem). The continuum limit of Eq. (10) should be considered as a short notation, as the system size expansion is valid only when the local population size $N_l = Na$ is large; this assumption clearly breaks down in the limit of $a \to 0$, see (Law et al., 2003; Doering et al., 2003; Hernandez-Garcia and Lopez, 2004; Birch and Young, 2006; Pigolotti et al., 2013) for examples in which this assumption is violated.

3 Model without total density conservation

We consider individuals as diffusing particles in d dimensions. We implement population dynamics by assuming that individuals of species i reproduce at rate μ_i and die with rates $\tilde{\lambda}_{ij}$ proportional to the number of individuals of species j in a given neighborhood. In a language borrowed from chemical kinetics, the "reactions" we consider are:

$$\begin{aligned} X_i &\overset{\mu_i}{\to} 2X_i \qquad \text{(reproduction)} \\ X_i + X_j &\overset{\tilde{\lambda}_{ij}}{\to} X_i \qquad \text{(death by competition)} \end{aligned} \tag{11}$$

In particular, competition occurs when individuals are within a small volume δ (for details on the numerical implementation of the individual-based dynamics see Perlekar et al. (2011)). We can then discretize the system in cells of size δ and start the derivation from the master equation governing the time evolution of the probability the numbers of particles $\{n_j^A, n_j^B\}$ of type A and B in each cell, labeled by the index j. We first define the rates $W_A(\pm 1, n_j^A, n_j^B)$ and $W_B(\pm 1, n_j^A, n_j^B)$ at which the populations of type A (or B) increase/decrease by one individual in a specific box, given that the population sizes are n_j^A and n_j^B. Letting aside the diffusion terms, the expression for these rates are:

$$\begin{aligned} W_A(+1, n_j^A, n_j^B) &= \mu_A n_j^A \\ W_A(-1, n_j^A, n_j^B) &= \tilde{\lambda}_{AA} n_j^A (n_j^A - 1) + \tilde{\lambda}_{AB} n_j^A n_j^B \\ W_B(+1, n_j^A, n_j^B) &= \mu_A n_j^B \\ W_B(-1, n_j^A, n_j^B) &= \tilde{\lambda}_{BA} n_j^A n_j^B + \tilde{\lambda}_{BB} n_j^B (n_j^B - 1). \end{aligned} \tag{12}$$

The master equation governing the evolution of the full probability distribution $P(\{n_j^A, n_j^B\}, t)$ for all possible box occupation numbers $\{n_j^A, n_j^B\}$

then reads:

$$
\begin{aligned}
\tfrac{d}{dt} & \; P(\{n_j^A, n_j^B\}, t) = \\
= & \sum_j [W_A(+1, n_j^A - 1, n_j^B) P(n_1^A, \ldots, n_j^A - 1, \ldots, n_1^B, \ldots) \\
- & \; W_A(+1, n_j^A, n_j^B) P(\{n_j^A, n_j^B\})] \\
+ & \sum_j [W_A(-1, n_j^A + 1, n_j^B) P(n_1^A, \ldots, n_j^A + 1, \ldots, n_1^B, \ldots) \\
- & \; W_A(-1, n_j^A, n_j^B) P(\{n_j^A, n_j^B\})] \\
+ & \sum_j [W_B(+1, n_j^A, n_j^B - 1) P(n_1^A, \ldots, n_1^B, \ldots, n_j^B - 1, \ldots) \\
- & \; W_B(+1, n_j^A, n_j^B) P(\{n_j^A, n_j^B\})] \\
+ & \sum_j [W_B(-1, n_j^A, n_j^B + 1) P(n_1^A, \ldots, n_1^B, \ldots, n_j^B + 1, \ldots) \\
- & \; W_B(-1, n_j^A, n_j^B) P(\{n_j^A, n_j^B\})] \\
+ & \; \text{diffusion terms,}
\end{aligned} \tag{13}
$$

where the diffusion terms account for the stochastic exchange of particles between neighboring boxes. As in the case of the stepping stone model, these terms reduce to discrete approximations to Laplace operator. Indeed, we will replace them with Laplacians in the continuous space limit at the end of the calculation.

In analogy with the previous section, we now to perform a Kramers-Moyal expansion (Risken, 1989) in each of the boxes. The only difference is that in this case it is a two-variable system, so we have to expand in the two independent increments Δn_A and Δn_B. The result is

$$
\begin{aligned}
\partial_t P\{n_j^A, n_j^B\} &= \sum_j \sum_{k=1}^{\infty} \frac{(-1)^k}{k!} \{\partial_{n_j^A}^k [\alpha_k^A(n_j^A, n_j^B) P(\{n_j^A, n_j^B\})] + \\
&+ \; \partial_{n_j^B}^k [\alpha_k^B(n_j^A, n_j^B) P(\{n_j^A, n_j^B\})]\},
\end{aligned} \tag{14}
$$

with the moments of the two jump distribution functions defined by

$$
\alpha_k^{A,B}(n_j^A, n_j^B) = \int d\Delta n_j^{A,B} \; (\Delta n_j^{A,B})^k \omega^{A,B}(\Delta n_j^{A,B}, n, j^A, n_j^B) \tag{15}
$$

and the function ω is defined from the rates exactly as in the previous section. Finally, truncating the Kramers-Moyal expansion up to second order in the derivatives leads to a Fokker-Planck equation for $P\{n_j^A, n_j^B\}$. It is convenient to write directly the equivalent but somewhat simpler system of Langevin equations corresponding to this Fokker-Planck description, namely:

$$\frac{dn_j^A}{dt} = n_j^A(\mu_A - \tilde{\lambda}_{AA}n_j^A - \tilde{\lambda}_{AB}n_j^B) + \text{diffusion} + \sigma_{A,j}\xi_j^A$$
$$\frac{dn_j^B}{dt} = n_j^B(\mu_B - \tilde{\lambda}_{BA}n_j^A - \tilde{\lambda}_{BB}n_j^B) + \text{diffusion} + \sigma_{B,j}\xi_j^B \quad (16)$$

where the noise amplitudes are

$$\begin{aligned} \sigma_{A,j}^2 &= n_j^A(\mu_A + \tilde{\lambda}_{AA}n_j^A + \tilde{\lambda}_{AB}n_j^B) \\ \sigma_{B,j}^2 &= n_j^B(\mu_B + \tilde{\lambda}_{BA}n_j^A + \tilde{\lambda}_{BB}n_j^B). \end{aligned} \quad (17)$$

In Eqns. (16), the ξ's are delta-correlated unit variance Gaussian processes, $< \xi_j^k(t)\xi_l^m(t') >= \delta_{jl}\delta_{km}\delta(t-t')$. In principle, the diffusion terms in (13) would contribute to the noise term. However, one can show that this contribution can be neglected if the size of the cells is sufficiently large (see Gardiner (2004)). In analogy with the previous section, from Eqs.(16) one can take (formally) the limit $\delta \to 0$. In such a way the number densities of individuals become continuous functions of the coordinate $\mathbf{x}$, $n_A(\mathbf{x}, t)$ and $n_B(\mathbf{x}, t)$.

We also define rescaled, macroscopic rates of binary reactions, $\lambda_{ij} = N\delta\tilde{\lambda}_{ij}$, and the macroscopic concentrations of individuals

$$c_{A,B}(\mathbf{x}, t) = n_{A,B}(\mathbf{x}, t)/N.$$

It is convenient to perform this rescaling in a different way for the well-mixed case (in which the population is not structured in space) and for the one dimensional case. In the former case we take $\delta = 1$. In analogy with the stepping stone model, calling $N_l = \delta N$ the local population size, we simply have $N = N_l$. In the spatial case, we fix $\delta = 1/N$ so that $\lambda_{ij} = \tilde{\lambda}_{ij}$, $\forall i, j$. This procedure leads to the following coupled spatial Langevin equations

$$\begin{aligned} \frac{\partial}{\partial t}c_A &= D\nabla^2 c_A + c_A(\mu_A - \lambda_{AA}c_A - \lambda_{AB}c_B) + \sigma_A\xi \\ \frac{\partial}{\partial t}c_B &= D\nabla^2 c_B + c_B(\mu_B - \lambda_{BA}c_A - \lambda_{BB}c_B) + \sigma_B\xi' \end{aligned} \quad (18)$$

where

$$\begin{aligned} \sigma_A^2 &= \frac{c_A(\mu_A + \lambda_{AA}c_A + \lambda_{AB}c_B)}{N} \\ \sigma_B^2 &= \frac{c_B(\mu_B + \lambda_{BA}c_A + \lambda_{BB}c_B)}{N}. \end{aligned} \quad (19)$$

4 Neutral theory

In this section, we present results in the simple case

$$\mu_A = \mu_B = \lambda_{AA} = \lambda_{AB} = \lambda_{BA} = \lambda_{BB}. \tag{20}$$

This case represent the neutral situation in which the two variants are neutral, i.e. phenotypically equivalent.

As before, we start our discussion with the well-mixed case. It is useful to describe the dynamics of the neutral version of the model in the c_A vs. c_B plane, represented in Fig. (1, left). Starting from a dilute initial condition, the system evolves rapidly towards to the intrinsic overall carrying capacity given by $c_A + c_B = 1$. The dynamics is then localized with fluctuations near this line, until extinction of one of the two species. This contrasts with the Moran process, in which the dynamics is rigidly confined to the $c_A + c_B = 1$ line. To assess the effect of these fluctuations, note from Eq. (18) that in the neutral case the total concentration $c_T = c_A + c_B$ obeys a closed equation:

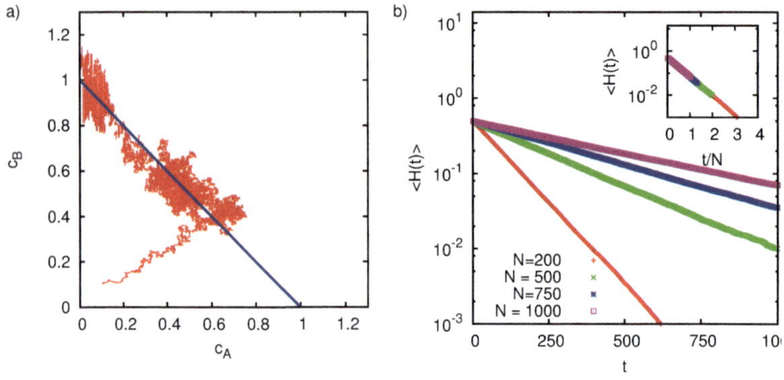

Figure 1. Neutral dynamics in the well-mixed case. (a) Example of a trajectory in the (c_A, c_B) plane with $N = 500$. The initial condition is $n_A = n_B = 20$, i.e. a small fraction of a typical long time carrying capacity. (b) Decay of the average heterozygosity $\langle H(t) \rangle$ for different values of N. Curves are obtained from simulations of the particle model; each curve is an average over 10^4 realizations and the error bars are smaller than the size of the lines. (inset) Collapse of the same curves plotted as a function of t/N. From Pigolotti et al. (2013).

$$\frac{d}{dt} c_T = \mu c_T (1 - c_T) + \sqrt{\frac{\mu c_T (1 + c_T)}{N}} \xi_c, \tag{21}$$

decoupled from the fraction of species A, $f = c_A/(c_A + c_B)$, where the noise term ξ_c satisfies $\langle \xi_c(t)\xi_c(t')\rangle = \delta(t-t')$. When N is large, the stationary solution, beside the solution $P(c) = \delta(c)$ corresponding to global extinction, is approximately a Gaussian with average $\langle c_T \rangle = 1$ and variance $\langle c_T^2 \rangle - \langle c_T \rangle^2 = N^{-1}$, which is small when N is large. We remind that, as in the particle model for simplicity death is implemented only via binary reactions (see Eq. 12), the state of global extinction is not accessible in the particle model, while it constitutes an absorbing state for Eq. 21. Such discrepancy with the macroscopic equation could be easily removed by allowing for death even in absence of competition, i.e. the reaction $X_i \to \emptyset$.

We now describe the dynamics of the relative fraction f of individuals carrying allele A, $f(t) = c_A/(c_A + c_B)$. Let us recall Ito's formula for a two variable system: let us write the Langevin equations for the two densities c_A and c_B as

$$\begin{aligned} \frac{d}{dt} c_A(\mathbf{x},t) &= \alpha_A(c_A, c_B) + \sigma_A(c_A, c_B)\xi(\mathbf{x},t) \\ \frac{d}{dt} c_B(\mathbf{x},t) &= \alpha_B(c_A, c_B) + \sigma_B(c_A, c_B)\xi'(\mathbf{x},t) \end{aligned} \tag{22}$$

where the diffusive Laplacian terms are included into α_A, α_B. The equation for $f(t)$ then reads

$$\begin{aligned} \frac{d}{dt} f &= \alpha_A \partial_A f + \alpha_B \partial_B f + \sqrt{\sigma_A^2 (\partial_A f)^2 + \sigma_B^2 (\partial_B f)^2}\,\xi + \\ &+ \frac{\sigma_A^2}{2} \partial_{AA} f + \frac{\sigma_B^2}{2} \partial_{BB} f, \end{aligned} \tag{23}$$

where we used the abbreviated notation $\partial_A = \partial_{c_A}$, $\partial_{AA} \equiv \partial^2_{c_A}$ and so on. Inserting the complete set of equations (18) into (23) leads to a lengthy expression for the dynamics of f. However, with the simple neutral choice of the parameters presented above in (20), the equation reduces to

$$\frac{d}{dt} f = \sqrt{\mu f(1-f)\frac{1 + c_T}{N c_T}}\,\xi_f \tag{24}$$

where $\xi_f(t)$ also satisfies $\langle \xi_f(t)\xi_f(t')\rangle = \delta(t-t')$, and further we have $\langle \xi_f(t)\xi_c(t')\rangle = 0$. The above equation is the same as the equation for the stepping stone model, Eq. (8), in the neutral case $s = 0$, apart from the coupling with the total density c_T which evolves dynamically according to (21). We can now analyze the global heterozygosity, which quantifies the loss of diversity as time evolves and is defined as the probability $H(t) = 2\langle f(1-f)\rangle$

that two randomly chosen individuals in the population carry different alleles. As the equation for c_T is independent of f in the neutral case studied here, one can factorize the average over c_T and f in the equation for $H(t)$:

$$\begin{aligned} \frac{d}{dt}H(t) &= -\frac{\mu}{N}\left\langle f(1-f)\frac{1+c_T}{c_T}\right\rangle = -\frac{\mu}{N}\langle f(1-f)\rangle\left\langle\frac{1+c_T}{c_T}\right\rangle \\ &= -\frac{2\mu}{N}H(t) + O\left(\frac{1}{N^2}\right). \end{aligned} \tag{25}$$

Neglecting the correction of order N^{-2}, we recover for our model with density fluctuations the closed equation for $H(t)$ for Fisher-Wright and Moran-type models with a fixed population size derived by Kimura, which states that the total heterozygosity decays exponentially in well mixed neutral systems (Crow and Kimura, 1970):

$$\langle H(t)\rangle = H(0)\exp(-2\mu t/N) \tag{26}$$

This exponential behavior is confirmed in simulations, as shown in Fig. (1b).

We now move to the one and two dimensional cases. To study how fixation occurs in space, we study the behavior of the *spatial* heterozygosity $H(x,t)$ defined as the probability of two individuals at distance x and time t to carry different alleles. In the neutral stepping stone model with a fixed population size in each deme, $H(x,t)$ obeys a closed equation:

$$\partial_t H(x,t) = 2D\nabla^2 H - \frac{2\mu}{N}H\delta(x). \tag{27}$$

In one dimension, such equation can be solved explicitly:

$$H(x,t) = H_0\left[1 - \frac{2}{N}\int_0^t dt' \frac{\text{erf}\left(\frac{t'}{4N^2D}\right)}{\sqrt{8\pi D(t-t')}}e^{-\frac{x^2}{8D(t-t')}+\frac{t'}{4N^2D}}\right] \tag{28}$$

where H_0 is the initial heterozygosity, equal to one half if the two variants are well mixed and equally populated at time $t=0$. Eqs. (27) and (28) can be derived directly from the stochastic Fisher equation (10) with $s=0$ (see, e.g., Korolev et al. (2009)).

We define the heterozygosity in the off-lattice particle simulations with growth and competition from the statistics of interparticle distances. In particular, at a given time t, we compute all distances between pairs of individuals. Upon introducing a bin size h, the function $H(r,t)$ is then defined as the ratio between the number of pairs carrying *different* alleles at a separation between r and $r+h$, divided by the total number of pairs of all types in the same range of separation. For simplicity, we always took the bin size h equal to the interaction distance δ.

In the limit $N\sqrt{D/\mu} \gg 1$, the spatial heterozygosity obtained by simulations of the neutral off-lattice model shows a remarkable agreement with Eq. (28), as shown in Fig. (2). This correspondence arises because, also in the spatial case, the relative fraction of allele A, $f(x,t) = c_A/(c_A + c_B)$, obeys a very similar equations as discussed in the mean field case. By applying Ito's formula in the spatial case as before, one can show that the only difference is an additional effective advection term in the equation for $\partial_t f$, equal to $2D(\nabla \log c_T) \cdot \nabla f$. The appearance of such terms was firstly found in Vlad et al. (2004) in a deterministic version of the model described here. Since c_T obeys a decoupled equation in the neutral case, such terms do not affect the equation for the heterozygosity. Indeed, numerical simulation shows that the average spatial heterozygosity in the model reproduces that of the stepping stone model even in the limit of very high diffusivity, as shown in Fig. 2, panel (b). Panel (c) shows that similar agreements arise comparing numerical integration of Eq. (27) with our off-lattice simulations in two dimensions. At variance with the one dimensional case, where the local heterozygosity $H(0,t)$ decays at long times as $t^{-1/2}$, in two dimension the decay is much slower, $H(0,t) \sim 1/\ln(t)$. Such slow logarithmic decay is confirmed in simulations in panel (d).

5 Conclusions

In this Chapter we compared two different stochastic models of spatially extended populations. We have shown that one can formally demonstrate their equivalence by means of stochastic calculus, at least in the case of neutral species. While the stepping stone model allows for a simpler analysis, the more general model is appropriate for cases in which the total density of individuals can vary considerably due to external causes, as a non-homogeneous distribution of resources or transport by fluid flows (Pigolotti et al., 2012, 2013).

Bibliography

D. A. Birch and W. A Young. A master equation for a spatial population model with pair interactions. *Theo. Pop. Biol.*, 70(1):2642, 2006.

J.F. Crow and M. Kimura. *Introduction to Population Genetics Theory.* Harper & Row Publishers, 1970.

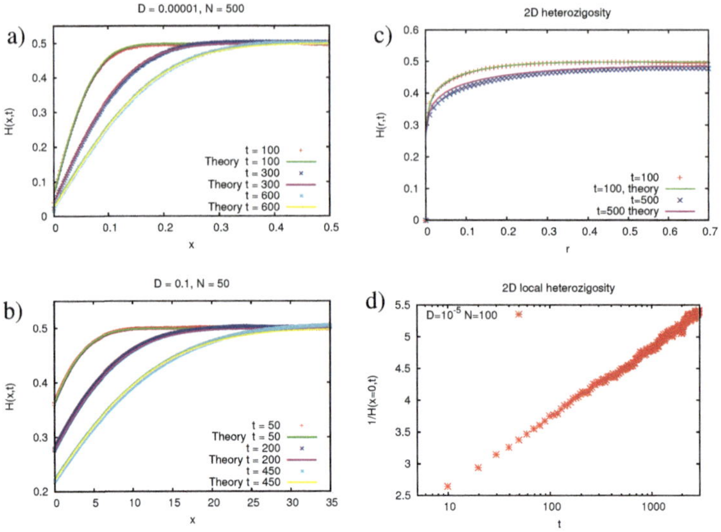

Figure 2. Heterozygosity in the 1d and 2d neutral case. Behavior of heterozygosity correlation function for the neutral off-lattice model of growth and competition. (a) 1D simulations at low diffusivity, $D = 10^{-5}$ and (b) high diffusivity, $D = 0.1$. In the top case, the system size is $L = 1$ while in the bottom case the system size is $L = 100$. In both cases we find excellent agreement with the prediction of formula (28). (c) Neutral heterozygosity in 2d, compared with a numerical integration of Eq. (27). (d) Behavior of the local heterozygosity $H(x = 0, t)$ as a function of time in 2D, showing the logarithmic decay $H(x = 0, t) \sim 1/\ln(t)$. From Pigolotti et al. (2013).

C. Doering, C. Mueller, and P. Smereka. Interacting particles, the stochastic fkpp equation, and duality. *Physica A*, 325:243–259, 2003.

C.W. Gardiner. *Handbook of Stochastic Methods.* Springer, 2004.

E. Hernandez-Garcia and C. Lopez. Clustering, advection and patterns in a model of population dynamics with neighborhood-dependent rates. *Phys. Rev. E*, 70(1):016216, 2004.

M. Kimura. "stepping stone" model of population. *Ann. Rept. Nat. Inst. Genetics*, 3:62–63, 1953.

M. Kimura and G. H. Weiss. The stepping stone model of population structure and the decrease of genetic correlation with distance. *Genetics*, 49: 561–576, 1964.

K.S. Korolev, M. Avlund, O. Hallatschek, and D.R. Nelson. Genetic demixing and evolutionary forces in the one-dimensional stepping stone model. *Review of Modern Physics*, 82:1691–1718, 2009.

R. Law, D. J. Murrell, and U. Dieckmann. Population growth in space and time: the spatial logistic equation. *Ecology*, 84(1):252–262, 2003.

J. D. Murray. *Mathematical Biology: an Introduction*. Springer, 2007.

P. Perlekar, R. Benzi, S. Pigolotti, and F. Toschi. Particle algorithms for population dynamics in flows. *Journal of Physics: Conference Series*, 333:012013, 2011.

S. Pigolotti, R. Benzi, M.H. Jensen, and D.R. Nelson. Population genetics in compressible flows. *Physical Review Letters*, 108:128102, 2012.

S. Pigolotti, R. Benzi, P. Perlekar, M.H. Jensen, F. Toschi, and D.R. Nelson. Growth, competition and cooperation in spatial population genetics. *Theoretical Population Biology*, 84:72–86, 2013.

H. Risken. *The Fokker-Planck equation: Methods of Solution and Applications*. Springer, Berlin, 1989.

M. Vlad, L. L. Cavalli-Sforza, and J. Ross. Enhanced (hydrodynamic) transport induced by population growth in reaction-diffusion systems with application to population genetics. *Proceedings of the National Academy of Sciences of the United States of America*, 101(28):10249–10253, 2004.

Discrete and Continuum Dynamics of Reacting and Interacting Individuals

Francesca Tesser * and Charles R. Doering †

* Department of Physics, Eindhoven University of Technology

† Department of Mathematics, Department of Physics, and Center for the Study of Complex Systems, University of Michigan, Ann Arbor

Abstract These lectures review some recent decades' research on non equilibrium statistical mechanics models of reaction and interaction kinetics. Lectures 1-4 focus on macroscopic kinetics of microscopically transport-limited interactions, while lectures 5 and 6 are concerned with discreteness and stochastic effects in reaction-diffusion fronts. The final lecture 7 considers demographic stochasticity in evolutionary population dynamics.

1 Macroscopic kinetics of microscopically transport-limited interactions

Traditional "mean field" or "mass action" reaction kinetic theories are extremely useful, but there are limits to their validity. In these lectures we examine the underlying assumptions that go into those approaches and consider situations where those assumptions are not valid. Several specific examples are developed to illustrate non-conventional reaction kinetics.

Microscopic interactions

One example of particle interactions is the electron-positron annihilation which gives as product two photons, $e^+ + e^- \to \gamma + \gamma$, and another example comes from chemistry, $2H + O \leftrightarrows H_2O$, where, in addition, we also include the reverse reaction. In general a two body reaction is a process in which two interacting particles A and B react to produce something as product:

$$A + B \to \text{Product}. \tag{1}$$

Sometimes the reactant particles are indistinguishable and the elementary interaction is

$$A + A \to \text{Product}. \tag{2}$$

While case (2) is a special version of the case (1), the two cases may behave quite differently dynamically. These two general prototypes collect a large number of systems in various fields: A and B could be an enzyme and a substrate respectively or a predator and a prey or else an infectious and a susceptible individual. The best way to describe the kinetics of these reactions is to trace back to the details of the interaction. This is possible only when the nature of the interaction is well-known, otherwise a classic mean field description is the first standard approach to that kind of systems.

Macroscopic kinetics

Often a macroscopic level of description is adopted for describing the kinetics of a general reaction $A + B \to C$ and $A + A \to C$, where C is the produced particle and for the moment no inverse reaction is considered, for example the particle C could be inert. The variables of the system are the mean field quantities $N_A(t)$, $N_B(t)$, $N_C(t)$, that are the total number of A, B and C particles in the system at time t. For the reactions of type $A + B$ the rate of variation of individuals of the A species is equal to that of the B and they are equal to the opposite variation of the C individuals

$$\frac{dN_A(t)}{dt} = \frac{dN_B(t)}{dt} = -\frac{dN_C(t)}{dt}.$$

For the reaction $A + A$ the following equality holds:

$$\frac{dN_A(t)}{dt} = -2\frac{dN_C(t)}{dt}.$$

The central question of reaction kinetics is what is the rate of variations of N_A and N_B?

In the mean field approach the Law of Mass Action asserts that the rate at which an elementary reaction proceeds is proportional to the product of the populations of the participating particles. In this theory the rate of reaction of the particles is proportional to the product of the numbers of reactants. For the $A + B$ reaction the mean field kinetics is

$$\frac{d}{dt}N_A = -kN_AN_B \tag{3}$$

introducing a phenomenological "rate constant" k. In the $A + A$ case the reaction rate is function of N_A only:

$$\frac{d}{dt}N_A = -kN_A^2. \tag{4}$$

Equations (3) and (4) describe the "reaction-limited" kinetics of the systems in which, roughly, each individual has opportunity to sample its neighborhood when it has a small probability of reaction per encounter. The rate of the reaction is then only related to the density of reactants neglecting any possible dependence on their spatial distribution. For initial conditions $N_A(t=0) = N_B(0) = N_0$ the solution for the ODE (3) is

$$N_A(t) = N_B(t) = \frac{N_0}{1 + N_0 kt}. \tag{5}$$

The exact same evolution in time holds for the equation (4) with initial condition $N_A(0) = N_0$. The long-time ($t \gg 1/(kN_0)$) dependence of the reactant particle numbers decays as t^{-1} in both systems, which we will refer to it as the mean field decay.

Transport-limited interactions

When the spatial distribution of the reactants becomes relevant for the macroscopic kinetics of the reaction other terms should take into account. We consider situations where the A and B particles are spatially distributed and proceed as random walkers. The proper variables for the system are now the local concentrations $a(\vec{x}, t)$ and $b(\vec{x}, t)$, and the diffusion process must be considered. At the macroscopic level the diffusion is often well described by the Fick's law where the flow of particles is proportional to the gradient of the concentration with proportionality constant D (which need not be the same for the two species, but for simplicity we limit consideration here to particles with equal diffusion coefficients). The temporal evolution of the $A+B$ reaction process is governed by the reaction-diffusion equations

$$\frac{\partial}{\partial t} a(\vec{x}, t) = D\nabla^2 a - \tilde{k}ab \tag{6}$$

$$\frac{\partial}{\partial t} b(\vec{x}, t) = D\nabla^2 b - \tilde{k}ab \tag{7}$$

where $\tilde{k}$ is the rate of the reaction per unit concentration. The first term represents the classical diffusion term, while the second term in the equations refers to the loss of particles due to reactions. For the single-species $A + A$ process the reaction-diffusion equation is

$$\frac{\partial}{\partial t} a(\vec{x}, t) = D\nabla^2 a - \tilde{k}a^2. \tag{8}$$

This sort of description neglects, by design, fluctuations and correlations which occur at microscopic level; more on this later.

In the case in which there is no flux of reactants at the boundary of the spatial domain and for initial concentrations $a(\vec{x},0) = b(\vec{x},0) = a_0 = b_0 = \rho_0$ the local concentrations $a(\vec{x},t)$ and $b(\vec{x},t)$ remain uniform in space because the diffusion term is zero so the concentrations obey

$$a(t) = b(t) = \frac{\rho_0}{1 + \rho_0 \tilde{k} t} \sim \frac{1}{\tilde{k} t} \qquad \text{as} \qquad t \to \infty. \tag{9}$$

For $t \gg 1/(\rho_0 \tilde{k})$, then, power law t^{-1} mean field decay is found.

If the condition of the flux is the same, but considering that the initial concentration is not homogeneous in space, but arbitrary, the total number of A or B particles in a volume Ω is the integral over the volume of the concentration $a(\vec{x},t)$ and $b(\vec{x},t)$ respectively:

$$N_A(t) = \int_\Omega a(\vec{x},t)d\vec{x} \qquad N_B(t) = \int_\Omega b(\vec{x},t)d\vec{x}.$$

More generally, for the single-species $A + A$ process with initial condition $a(\vec{x},0)$ and *any* sort of spatial variation in the concentration, $N_A(t)$ generally satisfies the inequality

$$N_A(t) \le \frac{N_0}{1 + \frac{N_0 \tilde{k} t}{|\Omega|}}$$

with $N_0 = \int_\Omega a(\vec{x},0)d\vec{x}$, displaying a mean-field-like *upper bound* on the decaying concentration, $\sim |\Omega| / \tilde{k} t$ independent of N_0 for large t.

Proof. For the $A+A$ process the concentration solves the Eq.(8). The total number of particles

$$N_A(t) = \int_\Omega a(\vec{x},t)d\vec{x}$$

evolves according to

$$\frac{d}{dt}N_A(t) = \int_\Omega \dot{a}(\vec{x},t)d\vec{x} = \int_\Omega (D\nabla^2 a(\vec{x},t) - \tilde{k} a(\vec{x},t)^2)d\vec{x} = -\tilde{k}\int_\Omega a(\vec{x},t)^2 d\vec{x}$$

since the diffusive term can be rewritten, thanks to the divergence theorem, as the integral over the boundaries of Ω and the term

$$\int_\Omega \nabla^2 a d\vec{x} = \int_{\partial\Omega} \vec{\nabla} a \cdot \hat{n} ds = 0$$

which vanishes when there is no flux of reactants at the boundary. Considering the general relation (the Cauchy-Schwarz inequality)

$$\left(\int_\Omega a\right)^2 \le |\Omega| \int_\Omega a^2,$$

we have

$$-\tilde{k}\int_{\Omega} a(\vec{x},t)^2 d\vec{x} \le -\frac{\tilde{k}}{|\Omega|}\left(\int_{\Omega} a(\vec{x},t)d\vec{x}\right)^2 = -\frac{\tilde{k}}{|\Omega|}N_A(t)^2.$$

Hence the total number of particles satisfies

$$\frac{d}{dt}N_A(t) \le -\frac{\tilde{k}}{|\Omega|}N_A(t)^2,$$

and for initial conditions $N_A(0) = N_0$ we can solve the differential inequality and deduce

$$N_A(t) \le \frac{N_0}{1 + N_0\frac{\tilde{k}t}{|\Omega|}}.$$

□

These considerations suggest that (a) spatial diffusion may play no role in macroscopic kinetics both in reaction-diffusion systems with homogeneous initial—and hence subsequent time-dependent—concentrations, and (b) for the irreversible single-species reaction, the mean field power law concentration decay $\sim t^{-1}$ is an upper limit in the presence of any sort of initial concentration variations. In the single-species case the implication is that spatial variations in concentration can only increase the bulk reaction rate.

The question we now address is whether these results are universally true. That is, we ask what aspects of the theory should change considering the underlying discrete nature of the interacting particle processes and the inevitable presence of spatial fluctuations in the system.

Macroscopic segregation for two-species reaction

In order to consider the effect of particle discreteness in the $A+B$ reaction system it is interesting to focus on the motion and behavior of single individuals. We consider statistically uniform initial distributions, independent and identical distributions, of A and B particles in a volume Ω, however now we note that even a statistically uniform distribution of particles contains fluctuations. In any sub-volume $V \subset \Omega$ containing N particles *on average*, there are statistical fluctuations on the order of the square root of $\sqrt{N}$. That is, if the initial condition is a statistically uniform random distribution of particles with density ρ_0, then

$$\int_V a(\vec{x},0) = \rho_0 V \pm \sqrt{\rho_0 V} \quad \text{and} \quad \int_V b(\vec{x},0) = \rho_0 V \pm \sqrt{\rho_0 V}.$$

Let us define $\rho(\vec{x},t)$ and $\gamma(\vec{x},t)$ as the sum and the difference of the two species' local concentrations:

$$\rho(\vec{x},t) = a(\vec{x},t) + b(\vec{x},t) \quad \text{and} \quad \gamma(\vec{x},t) = a(\vec{x},t) - b(\vec{x},t).$$

Equations (6) and (7) imply that ρ and γ satisfy

$$\partial_t \rho = D\nabla^2 \rho - \frac{1}{2}\tilde{k}\rho^2 + \frac{1}{2}\tilde{k}\gamma^2 \tag{10}$$

$$\partial_t \gamma = D\nabla^2 \gamma. \tag{11}$$

The equation for γ is the standard diffusion equation, linear in γ, while a term proportional to γ^2 appears in the equation for ρ acting as a *source* term for the total concentration in addition to the diffusion and the reaction terms.

The average of γ at time $t = 0$ is zero because of the statistically identical initial distribution of particles, $\langle\gamma(\vec{x},0)\rangle = 0$, so that the variance $\langle(\gamma - \langle\gamma\rangle)^2\rangle = \langle\gamma^2\rangle$. This can be computed because the independent initial distribution of particles are δ-correlated in space:

$$\langle\gamma(\vec{x},0)\gamma(\vec{x}',0)\rangle = 2\rho_0\delta(\vec{x}-\vec{x}').$$

This means that the variance of $\gamma(\cdot,0)$ integrated on every sub-volume V in d-spatial dimensions is

$$\begin{aligned}\left\langle\left(\int_V \gamma(\vec{x},0)d^dx\right)^2\right\rangle &= \left\langle\left(\int_V d^dx \int_V d^dx'\gamma(\vec{x},0)\gamma(\vec{x}',0)\right)\right\rangle \\ &= \int_V d^dx \int_V d^dx' \,\langle\gamma(\vec{x},0)\gamma(\vec{x}',0)\rangle \\ &= \int_V d^dx \int_V d^dx' 2\rho_0\delta(\vec{x}-\vec{x}') = 2\rho_0 \int_V d^dx = 2\rho_0 V\end{aligned}$$

so that, roughly speaking, the difference between the number of A and B particles in a sub volume V is initially

$$\int_V \gamma(\vec{x},0)d\vec{x} = 0 \pm \sqrt{2\rho_0 V}.$$

In order to calculate the contribution of $\gamma(\vec{x},t)^2$ in equation (10) it is convenient to compute the Fourier transform

$$\hat{\gamma}(\vec{k},t) = \frac{1}{(2\pi)^d}\int_\Omega e^{-i\vec{k}\cdot\vec{x}}\gamma(\vec{x},t)d^dx.$$

The correlation of $\hat{\gamma}(\vec{k},0)$ is

$$
\begin{aligned}
\left\langle\gamma(\vec{k},0)\gamma(\vec{k}',0)\right\rangle &= \frac{1}{(2\pi)^{2d}}\int_\Omega d^dx\int_\Omega d^dx' e^{-i(\vec{k}\cdot\vec{x}+\vec{k}'\cdot\vec{x}')}\left\langle\gamma(\vec{x},0)\gamma(\vec{x}',0)\right\rangle \\
&= \frac{1}{(2\pi)^{2d}}\int_\Omega d^dx\int_\Omega d^dx' e^{-i(\vec{k}\cdot\vec{x}+\vec{k}'\cdot\vec{x}')}2\rho_0\delta(\vec{x}-\vec{x}') \\
&= \frac{2\rho_0}{(2\pi)^{2d}}\int_\Omega d^dx e^{-i(\vec{k}\cdot\vec{x}+\vec{k}'\cdot\vec{x})} = \frac{2\rho_0}{(2\pi)^d}\delta(\vec{k}+\vec{k}')
\end{aligned}
$$

where the last passage is relevant to the limit of large Ω. Since it is linear, the Fourier transform of the partial differential equation (11) gives the ordinary differential equation

$$
\frac{d}{dt}\hat{\gamma}(\vec{k},t) = -Dk^2\hat{\gamma}(\vec{k},t) \tag{12}
$$

with solution

$$
\hat{\gamma}(\vec{k},t) = \hat{\gamma}(\vec{k},0)e^{-Dk^2t}. \tag{13}
$$

The correlation of $\hat{\gamma}$ at time t is given by

$$
\left\langle\gamma(\vec{k},t)\gamma(\vec{k}',t)\right\rangle = \left\langle\gamma(\vec{k},0)\gamma(\vec{k}',0)\right\rangle e^{-D(k^2+k'^2)t} = \frac{2\rho_0}{(2\pi)^d}\delta(\vec{k}+\vec{k}')e^{-2Dk^2t},
$$

so the ensemble average integrated over the volume Ω of γ^2 is

$$
\begin{aligned}
\left\langle\left(\int_\Omega \gamma(\vec{x},t)^2 d^dx\right)\right\rangle &= \left\langle\left(\int_\Omega d^dx\gamma(\vec{x},t)\gamma(\vec{x}',t)\right)\right\rangle \\
&= \left\langle\int_\Omega d^dx\int_\Omega d^dk e^{i\vec{k}\cdot\vec{x}}\hat{\gamma}(\vec{k},t)\int_\Omega d^dk' e^{i\vec{k}'\cdot\vec{x}}\hat{\gamma}(\vec{k}',t)\right\rangle \\
&= \int_\Omega d^dx\int_\Omega d^dk\int_\Omega d^dk' e^{i(\vec{k}+\vec{k}')\cdot\vec{x}}\frac{2\rho_0}{(2\pi)^d}\delta(\vec{k}+\vec{k}')e^{-2Dk^2t} \\
&= |\Omega|\frac{2\rho_0}{(2\pi)^d}\int_\Omega d^dk e^{-2Dk^2t} \\
&= |\Omega|\frac{2\rho_0}{(2\pi)^d}dC_d\int_0^\infty dk k^{d-1}e^{-2Dk^2t}
\end{aligned}
$$

where dC_dk^{d-1} is the surface of the d-1 sphere of ray k. With the change in variable $\xi = \sqrt{2Dt}k$ it is possible to recognize the Gamma function $\Gamma(d/2)$:

$$
\begin{aligned}
&= |\Omega|\frac{2\rho_0}{(2\pi)^d}dC_d\frac{1}{(2Dt)^{d/2}}\int_0^\infty d\xi\xi^{d-1}e^{-\xi^2} \\
&= |\Omega|\frac{\rho_0}{(2\pi)^d}dC_d\frac{\Gamma(\frac{d}{2})}{(2Dt)^{d/2}}.
\end{aligned}
$$

The evolution in time for that quantity at time gives a power law in time which depends on the spatial dimension d:

$$\left\langle \int_\Omega \gamma(\vec{x},t)^2 d\vec{x} \right\rangle \sim \rho_0 \, |\Omega| \, (Dt)^{-\frac{d}{2}}. \tag{14}$$

Thus, even when the diffusion term does not influence the bulk dynamics because of uniform average initial conditions the equation for ρ Eq. (10) is made of two terms

$$\begin{aligned} \frac{d}{dt} \int_\Omega \rho(\vec{x},t) d\vec{x} &= -\frac{1}{2}\tilde{k} \int_\Omega \rho(\vec{x},t)^2 d\vec{x} + \frac{1}{2}\tilde{k} \int_\Omega \gamma(\vec{x},t)^2 d\vec{x} \\ &\approx -\frac{1}{2}\frac{\tilde{k}}{|\Omega|} \left[\int_\Omega \rho(\vec{x},t) d\vec{x} \right]^2 + \mathcal{O}\left(\tilde{k}\rho_0 \, |\Omega| \, (Dt)^{-\frac{d}{2}} \right). \end{aligned}$$

The error committed in moving the square outside the first integral will be negligible when the *total* density of reactants ρ is spatially uniform—even though the individual densities a and b may not be. The balance of the two terms depends on the spatial dimension: the second term—balancing the first term—yields a $t^{-\frac{d}{4}}$ decay which dominates mean field t^{-1} for $d < 4$. The conclusion is that the spatial dimension is essential for considering the kinetics of the number of particles in such systems:

$$N_A(t) = N_B(t) = \frac{1}{2} \int_\Omega \rho(\vec{x},t) d\vec{x} \sim \begin{cases} \sqrt{|\Omega| \, N_0} (Dt)^{-d/4} & \text{for} \quad d \leq 4 \\ |\Omega| \, (\tilde{k}t)^{-1} & \text{for} \quad d \geq 4. \end{cases}$$

The main result is the appearance of anomalous (non-mean-field) scaling for $d < 4$.

The effect of fluctuations like this was first discussed by Ovchinnikov and Zeldovich (1978). They analyzed the influence of initial density fluctuations of the reactants of the $A + B$ process in 3-dimensional volume, obtaining the more slow time dependance $t^{-3/4}$ instead of t^{-1} for the evolution of the number of particles in time. Later, numerical and analytical work (Toussaint and Wilczek, 1983) was performed for the same process of particles and antiparticles moving diffusively and annihilating irreversibly. For a two dimensional system they studied the time decay for the density of particles changing the lattice size for the domain and they investigated the most dramatic macroscopic manifestation of the anomalous decay for d smaller than 4: the spontaneous macroscopic segregation of reactants. That is, the emergence of regions of only one type of particles. These islands are spontaneously created because the diffusion is not fast enough to dissolve them. In those configurations reactions only occur at interfaces of subdomains.

Macroscopic segregation was proved to appear also in fractal dimension (Kopelman, 1988), the $A + B$ reaction simulated in a Sierpinski triangle gives an anomalous decay for the density of particles $\rho \approx t^{-0.4}$ for $d \sim 1.6$.

A rigorous proof of the anomalous time scaling was presented in the paper by Bramson and Lebowitz (1991), while the first experimental demonstration of the spontaneous segregation in three dimensions was provided in 2000 (Monson and Kopelman, 2000). In order to postpone the reaction until equilibrium random initial conditions had been achieved, Monson and Kopelman encapsulated one of the two species and used laser speckles to start the process. The system obtained by relaxing the restriction to irreversible reactions, i.e., the process which allows the back reaction $A + B \leftrightarrows C$, exhibits its own peculiarities (ben Avraham and Doering, 1988; Clément et al., 1989).

In conclusion, these examples show that transport properties of individual discrete reactants play a great role in the kinetics of reactions. This fact is most emphasized when reactions occur in very geometrical complicated environment; indeed, it has much more recently been appreciated how much the first-passage time (the random time for a molecule to reach a target) is related both to the initial distance between reactants and complex geometric constraints (Bénichou et al., 2010).

Microscopic segregation for single-species reaction

In the previous section, a macroscopic segregation phenomenon for two species diffusion-limited reactions was described. Now we would like to consider processes where only one species is present. For this kind of processes no macroscopic segregation is possible, but anomalous scaling still occurs. In the paper of Toussaint and Wilczek (1983), a scaling argument for single-species reactions suggested

$$N_A(t) = \int_\Omega \rho(\vec{x},t)d\vec{x} \sim \begin{cases} |\Omega|\,(Dt)^{-d/2} & \text{for} \quad d \leq 2 \\ |\Omega|\,(\tilde{k}t)^{-1} & \text{for} \quad d \geq 2. \end{cases} \tag{15}$$

That is,

$$\frac{d}{dt}N_A(t) \sim \begin{cases} -N_A^{\frac{d+2}{d}} & \text{for} \quad d \leq 2 \\ -N_A^2 & \text{for} \quad d \geq 2. \end{cases} \tag{16}$$

This means that mean field behavior for the evolution of the number of particles (or the evolution for the density) holds for sufficiently high spatial dimensions d, but not so in low dimensions. The goal now is to develop an exact solution when the mean field kinetics does not hold. Three different processes are discussed in this section:

1. the coagulation process $A + A \to A$;
2. the process $A + A \to A$ together with a source, $\emptyset \to A$;
3. the reversible process $A + A \leftrightarrows A$.

Mean field mass action theory gives the following kinetic equations, respectively, for the concentration in these three processes:

1. $d\rho/dt = -\tilde{k}\rho^2$
2. $d\rho/dt = -\tilde{k}\rho^2 + R$
3. $d\rho/dt = -\tilde{k}\rho^2 + \beta\rho$,

where $\tilde{k}$ is the coagulation rate, β the birth rate and R is the rate of creation. We note that the last equation is the logistic equation.

Now, consider the processes in the limit where the characteristic time for reaction, given two particles in each others' neighborhoods, is much smaller than the characteristic time for diffusion of the particles into each others' neighborhoods. This is when the effect of "self stirring" by diffusion is ineffective and the law of mass action does not hold. The associated diffusion-limited kinetics may very well be different.

As a first guess one can assume the validity of the scaling argument above and infer from equations (16) that the diffusion-limited kinetics in the $d = 1$ case for the three processes are

1. $d\rho/dt \sim -D\rho^3$
2. $d\rho/dt \sim -D\rho^3 + R$
3. $d\rho/dt \sim -D\rho^3 + \beta\rho$.

As will be shown, this approach fails to describe the real dynamics in all these. In the following we will extract the exact solution from a microscopic treatment (ben Avraham et al., 1990) in order to have a direct comparison with the mean field description.

The one-species process we consider is the transport-limited reversible coagulation process $A + A \leftrightarrows A$ with input $\emptyset \to A$ and diffusion on a one dimensional lattice where Δx is the space between adjacent sites. Each site can be singly occupied or empty. The diffusion process is modeled giving a hop rate for a particle to a neighboring site of $D(x,t)/\Delta x^2$, where $D(x,t)$ is the (local) macroscopic diffusion coefficient which may be space (x) and time dependent. The birth process is described in this way: a particle at site x gives birth to another particle at site $x \pm \Delta x$ at rate $v(x,t)/2\Delta x$. The coagulation/coalescence phenomenon is enforced by fixing the number of occupation at each sites 0 or 1, so that two particles on the same site simply become one. To complete the scenario the spontaneous generation of particles can occur with probability at each site equal to $R(x,t)\Delta x$ (R is a probability per unit length, which may be space and time dependent).

An interesting quantity is the empty interval probability $E(x, y, t)$ defined as the probability that all the sites between x and y (including x and y) are empty at time t:

$$E(x, y, t) = Prob\,\{\text{sites from } x \text{ to } y \geq x \text{ are empty at time t}\}$$

$= Prob\{$ 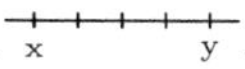 $\}.$

If $\rho(x, t)$ is the density of particles at site x and time t,

In order to derive the kinetic equation for E we need an expression for the variation of E, ΔE in time interval short enough to neglect double births, hops, etc. The combination of all processes in time interval Δt produce an increasing of $E(x, y)$ in the cases where there is only one particle in x and it hops to the left or only one particle in y that hops to the right. On the contrary there is a decrease of $E(x, y)$ in the cases where a particle in $x - \Delta x$ hops to x, one particle in $y + \Delta x$ hops to y, a particle in $x - \Delta x$ gives birth to a particle into site x, a particle in $y + \Delta x$ gives birth to a particle in y or in cases where there is an input of particles at each site z between x and y. Altogether we may write

$\Delta E = \dfrac{D(x,t)\Delta t}{\Delta x^2} Prob\{$  $\} + \dfrac{D(y,t)\Delta t}{\Delta x^2} Prob\{$ 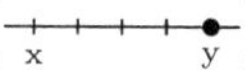 $\}$

$- \dfrac{D(x - \Delta x, t)\Delta t}{\Delta x^2} Prob\{$ 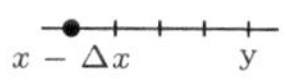 $\}$

$- \dfrac{D(y + \Delta x, t)\Delta t}{\Delta x^2} Prob\{$ 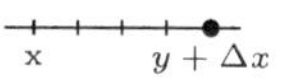 $\}$

$- \dfrac{v(x - \Delta x, t)\Delta t}{\Delta x} Prob\{$ 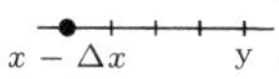 $\}$

$- \dfrac{v(y + \Delta x, t)\Delta t}{\Delta x} Prob\{$ 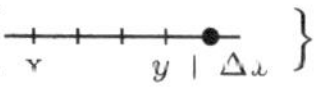 $\}$

$- [R(x, t) + R(x + \Delta x, t) + \cdots + R(y, t)]\, \Delta x E(x, y, t).$

The closure of the kinetic equation comes from the ability to write each configuration as a union of two others configurations, for example:

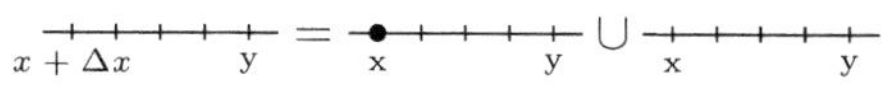

so that the corresponding probability can be written as the sum of two probabilities:

$E(x + \Delta x, y) = Prob\{$ 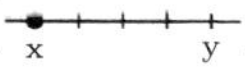 $\} + E(x, y)$

with analogous passages:

$$E(x, y - \Delta x) = Prob\{$$ 

$$\} + E(x, y)$$

$$E(x, y) = Prob\{$$ 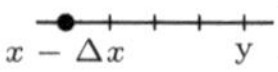

$$\} + E(x - \Delta x, y)$$

$$E(x, y) = Prob\{$$ 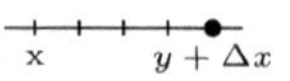

$$\} + E(x, y + \Delta x).$$

These expressions can be substituted in the equation for ΔE, then dividing by Δt and considering the limit $\Delta t \to 0$ we obtain a differential equation for E:

$$\begin{aligned}
\frac{dE(x,y,t)}{dt} =& \frac{D(x,t)}{\Delta x^2}\{E(x+\Delta x, y, t) - E(x,y,t)\} \\
&+ \frac{D(y,t)}{\Delta x^2}\{E(x, y-\Delta x, t) - E(x,y,t)\} \\
&- \frac{D(x-\Delta x,t)}{\Delta x^2}\{E(x,y,t) - E(x-\Delta x, y, t)\} \\
&- \frac{D(y+\Delta x,t)}{\Delta x^2}\{E(x,y,t) - E(x, y+\Delta x, t)\} \\
&- \frac{v(x-\Delta x)}{2\Delta x^2}\{E(x,y,t) - E(x-\Delta x, y, t)\} \\
&- \frac{v(y+\Delta x)}{2\Delta x^2}\{E(x,y,t) - E(x, y+\Delta x, t)\} \\
&- \sum_{z=x}^{y} R(z,t)\Delta x E(x,y,t).
\end{aligned} \tag{17}$$

In order to make the equation valid for $x = y$ the appropriate boundary conditions are

$$E(x+\Delta x, x, t) = 1 \qquad \text{and} \qquad E(x, x-\Delta x, t) = 1.$$

In the spatial continuum limit $\Delta x \to 0$ we find the partial differential equation

$$\begin{aligned}
\frac{\partial E(x,y,t)}{\partial t} =& \frac{\partial}{\partial x}\left[D(x,t)\frac{\partial E}{\partial x}\right] + \frac{\partial}{\partial y}\left[D(y,t)\frac{\partial E}{\partial y}\right] - \frac{v(x,t)}{2}\frac{\partial E}{\partial x} \\
&+ \frac{v(y,t)}{2}\frac{\partial E}{\partial y} - \left\{\int_x^y R(z,t)dz\right\} E(x,y,t)
\end{aligned} \tag{18}$$

which is linear in E, with boundary conditions

$$\lim_{y \to x^+} E(x,y,t) = 1 \qquad \text{and} \qquad \lim_{x \to y^-} E(x,y,t) = 1.$$

In the continuum limit $\Delta x \to 0$ the statistical local density

$$\rho(x,t) = \frac{1}{\Delta x}\left[1 - E(x,x,t)\right] = \frac{1}{\Delta x}\left[E(x,x-\Delta x,t) - E(x,x,t)\right] \qquad (19)$$

is given by

$$\rho(x,t) = -\left.\frac{\partial E(x,y,t)}{\partial y}\right|_{y=x}. \qquad (20)$$

In the case of constant D, R and v, Eq.(18) becomes a constant coefficient PDE

$$\frac{\partial E(x,y,t)}{\partial t} = D\left(\frac{\partial^2 E}{\partial x^2}\right) + D\left(\frac{\partial^2 E}{\partial y^2}\right) - \frac{v}{2}\frac{\partial E}{\partial x} + \frac{v}{2}\frac{\partial E}{\partial y} - R(y-x)E(x,y,t), \qquad (21)$$

valid in the half-space $y \geq x$. Proceeding with the change in variables $\xi = y + x$ and $\zeta = y - x$, the PDE (21) becomes:

$$\frac{\partial E(\xi,\zeta,t)}{\partial t} = 2D\frac{\partial^2 E(\xi,\zeta,t)}{\partial \xi^2} + 2D\frac{\partial^2 E(\xi,\zeta,t)}{\partial \zeta^2} + v\frac{\partial E}{\partial \zeta} - R\zeta E(\xi,\zeta,t). \qquad (22)$$

But $\partial/\partial\xi = 0$ in (statistically) translational invariant situations where E is only a function of ζ and t and (22) reduces to

$$\frac{\partial E(\zeta,t)}{\partial t} = 2D\frac{\partial^2 E(\zeta,t)}{\partial \zeta^2} + v\frac{\partial E}{\partial \zeta} - R\zeta E(\zeta,t) \qquad (23)$$

with boundary conditions

$$E(0,t) = 1 \qquad \text{and} \qquad E(\infty,t) = 0.$$

In this case the expression for the density is

$$\rho(t) = -\left.\frac{\partial E}{\partial \zeta}\right|_{\zeta=0}. \qquad (24)$$

In order to find a general solution for the Eq.(23) we may expand E as linear superposition of eigenfunctions

$$E(\zeta,t) = \sum_{\lambda} a_\lambda E_\lambda(\zeta) e^{-\lambda t}$$

where

$$-\lambda E_\lambda(\zeta) = 2D\frac{\partial^2 E_\lambda(\zeta)}{\partial \zeta^2} + v\frac{\partial E_\lambda(\zeta)}{\partial \zeta} - R\zeta E_\lambda(\zeta). \qquad (25)$$

Now let $E_\lambda(\zeta) = F_\lambda(\zeta)e^{-\frac{v\zeta}{4D}}$ where

$$2D\frac{\partial^2 F_\lambda(\zeta)}{\partial\zeta^2} = \left[R\zeta + \left(\frac{v^2}{8D} - \lambda\right)\right] F_\lambda(\zeta). \tag{26}$$

This is Airy's equation with solution

$$F_\lambda(\zeta) = Ai\left\{\left(\frac{R}{2D}\right)^{\frac{1}{3}} \zeta + \left(\frac{v^2}{8D} - \lambda\right)\frac{1}{(2DR^2)^{\frac{1}{3}}}\right\}. \tag{27}$$

As $z \to \infty$, $Ai(z) \sim e^{-z^{3/2}}$. The steady state solution is obtained from Eq.(27) setting $\lambda = 0$, whereas the transient solutions correspond to $\lambda > 0$ and, combining with the boundary conditions, $F_\lambda(0) = 0$ for $\lambda > 0$. Applying this to Eq.(27) it is possible to obtain a discrete relaxation spectrum for nonvanishing R and D:

$$\lambda_n = \frac{v^2}{8D} + (2DR^2)^{\frac{1}{3}} |a_n| \tag{28}$$

where a_n is the nth zero of the Airy function $Ai(z)$. These zero are all negative and their values are tabulated in the literature.

Let us now analyze the three processes introduced at the beginning of the section. The first case presented was the irreversible coagulation process $A + A \to A$ corresponding to $R = 0$ and $v = 0$. The steady state is a trivial solution with zero concentration of particles, instead the kinetic solution for the density of particles gives at large time the non-mean-field behavior:

$$\rho(t) \to \frac{1}{\sqrt{2\pi Dt}} \quad \text{as} \quad t \to \infty \quad \text{and} \quad \frac{d\rho}{dt} = -\pi D\rho^3. \tag{29}$$

In an experimental work Kroon et al. (1993) study the kinetics of the tetramethylammonium magnese trichloride (TMMC) which turns out to be a perfect model for diffusion-reaction system in 1D. A full characterization of the coalescence process is provided only by the infinite hierarchy of correlation functions and a complete exact description is given in the paper of ben Avraham (1998).

In the second class of coalescence process also an input is present, so $v = 0$, but $R > 0$. The dynamics of the solution for the density near the steady state, forced onto a rate equation form, must be written

$$\frac{d\rho}{dt} \approx -c_1 D\rho^3 + c_2 R \tag{30}$$

where

$$c_1 = \frac{2\,|a_1|\,Ai(0)^2}{3Ai'(0)^2} \neq \pi \qquad \text{and} \qquad c_2 = \frac{|a_1| \cdot |Ai'(0)|}{3Ai(0)} \neq 1.$$

These are both different from the mean field behavior *and* from what one might guess by adding the source term to the decay dynamics in (29).

In the last case we analyze the process with both forward and backward reaction, so $v > 0$, but with no input $R = 0$. Near the equilibrium statistically steady state, logistic dynamics is found:

$$\frac{d\rho(t)}{dt} \approx \beta\rho(t)\left(1 - \frac{\rho(t)}{\rho_{eq}}\right) \tag{31}$$

with $\beta = \frac{1}{2}D\rho_{eq}^2$. The stationary state of this reversible process is a true thermodynamic equilibrium with a totally random (Poisson) distribution of particles.

But the approach to equilibrium can be slower than that suggested by logistic approach for far-from-equilibrium initial conditions. Unlike the predictions of any mean field rate equations, the asymptotic relaxation rate depends on initial conditions: this process exhibits a sharp transition in its relaxation dynamics. Considering initial conditions consisting of purely random distributions of particles with density ρ_0 and writing the long-time density $\rho(t) = \rho_{eq} + O(e^{-t/\tau_r})$, the relaxation time τ_r is found to obey

$$\tau_r = \begin{cases} \beta^{-1} & \text{for} \quad \rho_0 \geq \frac{1}{2}\rho_{eq} \\ [2D\rho_0(\rho_{eq} - \rho_0)]^{-1} & \text{for} \quad \rho_0 \leq \frac{1}{2}\rho_{eq}. \end{cases} \tag{32}$$

This kinetic phase transition is the result of long-lived spatial correlations in the microscopic distribution of particles appearing when initial state is far enough from equilibrium. The spatial correlations between particles positions persist forever when $\rho_0 < \frac{1}{2}\rho_{eq}$ and the kinetic phase transition is destroyed with finite volume (Doering and Burschka, 1990).

The mechanism for the appearance of the "slow-relaxation" phase is clear when the initial density is much smaller that the equilibrium density. When $\rho_0 \ll \rho_{eq}$, the typical distance between particles, $l \sim 1/\rho_0$, is large and a concentration wavefront must propagate in order to fill the gaps between the particles. The speed of the front is $v/2$ (Doering et al., 1991) which results in the "long" time scale $\sim (\rho_0 v)^{-1}$. We note that the average concentration front in one dimension never reaches a unique shape but it appears a continuous spreading due to diffusion of the leading particle although in each realization the shape of the wave remain actually constant (ben Avraham, 1998).

In conclusion, we see that microscopic fluctuations are relevant for the kinetics of diffusion-limited processes and can spontaneously produce correlations among participants that diffusive transport may not be able to dissipate. In case of two species reacting ($A + B$) macroscopic correlations may emerge producing anomalous kinetics even though local mean field theory may still apply. On the other hand the single species ($A + A$) reactions display anomalous kinetics due to microscopic segregation and even possess a macroscopic phase transition which depends explicitly on long-lived microscopic spatial correlations. These examples illustrate difficulty of any kind of general macroscopic kinetic theory and, in particular, the failure of law of mass description of the dynamics.

2 Discreteness and stochastic effects in reaction-diffusion fronts

In these two lectures we develop an example illustrating how microscopic discreteness and fluctuations can qualitatively modify macroscopic dynamics, in this case the propagation of a reaction-diffusion front. The analysis involves developing a correspondence between two different reaction processes, one of which is naturally described in terms of a nonlinear stochastic partial differential equation with multiplicative noise. This discussion comes from Doering et al. (2003).

The stochastic Fisher-Kolmogorov-Petrovsky-Piscounov equation

The stochastic FKPP equation is

$$\partial_t U(x,t) = D\partial_{xx}U + \gamma U(1-U) + \epsilon\sqrt{U(1-U)}\,\eta(x,t) \tag{33}$$

where $0 \leq U(x,t) \leq 1$ and $\eta(x,t)$ a Gaussian white noise. Solutions of this stochastic partial differential equation have a special connection with the reaction-diffusion process $A \leftrightarrows A + A$. This "duality" relationship is an exact mathematical connection between the particle process with appropriate growth and diffusion rates and the solutions $U(x,t)$ of the equation, so it is a useful tool for transferring some results among them. We begin by motivating the stochastic partial differential equation (33).

The logistic ordinary differential equation

$$\frac{d}{dt}U(t) = \gamma U(1-U) \tag{34}$$

is characterized by the growth rate γ and a saturation term (here normalized to 1). This equation has two stationary solutions, one is the unstable solution for $U = 0$ and the other is the stable fixed point $U = 1$ corresponding to the saturation value. In 1937 Fisher and independently Kolmogorov, Petrovsky and Piscounov studied the addition of the spatial dependence in the logistic system through the insertion of the diffusive term obtaining a description of the macroscopic spatial spreading of a gene in "empty" space when interpreting $U(x,t)$ as the local density of the population at time t. The reaction-diffusion equation

$$\frac{\partial}{\partial t}U(x,t) = D\frac{\partial^2 u}{\partial x^2} + \gamma U(1-U) \tag{35}$$

is thus known as the FKPP equation. It is characterized by the diffusion coefficient D in addition to the growth rate γ. The FKPP equation with initial conditions satisfying

$$\lim_{x\to-\infty} U(x,t) = 1 \qquad \text{and} \qquad \lim_{x\to+\infty} U(x,t) = 0$$

admits traveling wave solutions, that is to say solutions of the form

$$U(x,t) = F(z) = F(x-ct). \tag{36}$$

This corresponds to a rigid translation of a front with speed c in positive x direction, where c is to be determined.

The insertion of the wave solution ansatz (36) into (35) gives the ordinary differential equation

$$DF''(z) + cF'(z) + \gamma F(z)(1-F(z)) = 0 \tag{37}$$

together with initial conditions

$$\lim_{z\to-\infty} F(z) = 1 \qquad \text{and} \qquad \lim_{z\to+\infty} F(z) = 0.$$

This ODE (37) can be studied as the mechanical analog of the dynamics of a body of convenient mass moving in the potential

$$V(F) = \frac{1}{2}F^2 - \frac{1}{3}F^3$$

with velocity-dependent friction characterized by the coefficient c; see Figure 2. The solutions divide in different classes, depending on the strength of the friction.

If c is low enough the mass joins $F = 0$ after oscillations around the equilibrium. In case of high values of c no oscillations are present, and $c_{min} = 2\sqrt{D\gamma}$ is the minimum value for c corresponding to a solution which join the stable fixed point without crossing it. In order to keep its physical meaning as a population or concentration, $U(x,t)$ must be non-negative so the oscillating solutions are not acceptable. This excludes all possible values under the critical speed c_{min} while all values above it are accessible to the system, including the critical value itself. Moreover if suitable initial conditions are assumed, i.e., when the front is sharp enough, a weak velocity selection holds and the final speed of the front converges to the minimum value $c \sim c_{min} + \mathcal{O}(t^{-1})$ (Bramson, 1983 and Ebert and van Saarloos, 2000). This slow speed selection by propagation of a stable phase into the unstable phase presents a fundamental problem because it depends on the specific shape of the asymptotic structure of the leading tail, where, additionally, the model is most unstable. This makes one wonder how robust the result is to physically meaningful perturbations.

Discrete nature of the microscopic process

The microscopic approach gives a better understanding of the system because it considers intrinsically the discreteness nature of particles and fluctuations. In this section it is shown how it is possible to include these features of the system in the continuum logistic kinetics trough the addition of a noise term. Consider the $A + B$ reaction scheme:

$$A + B \xrightarrow{k_1} A + A$$

$$A + B \xrightarrow{k_2} B + B$$

in a well stirred reactor of volume Ω all A particles have chance to react with B and vice-versa with rates k_1 and k_2. $N_A(t)$ and $N_B(t)$ are the number of A and respectively B particles at time t and the total number of particles $N = N_A(t) + N_B(t)$ is conserved by these reactions.

A natural variable is $U(t) = N_A(t)/N$ and the mean field equation for its kinetics is

$$\frac{d}{dt}U(t) = \left(\frac{k_1 - k_2}{\Omega}\right) U(1 - U), \tag{38}$$

the logistic equation with $\gamma = (k_1 - k_2)/\Omega$. On other hand $N_A(t)$ is really a discrete microscopic Markov process with $p_n(t) = \text{Prob}(N_A(t) = n)$, $n = 0, 1, \ldots, N$, evolving according to the master equation

$$\begin{aligned}\frac{dp_n}{dt} = &- \frac{k_1}{\Omega} n(N-n)p_n + \frac{k_1}{\Omega}(n-1)(N-n+1)p_{n-1} \\ &- \frac{k_2}{\Omega} n(N-n)p_n + \frac{k_2}{\Omega}(n+1)(N-n-1)p_{n+1}.\end{aligned} \tag{39}$$

Here it is convenient to take $p_{-1} = 0 = p_{N+1}$ in order to apply the master equation also at extreme values $n = 0$ and $n = N$.

Changing the variable to $u = n/N$ with $f(u,t) = Np_{Nu}(t)$ which formally converges to the probability density function for u, discreteness is represented by $\Delta u = 1/N$. After a Taylor expansion in terms of $u+\Delta u$ and $u - \Delta u$, in the limit of continuum approximation $N \gg 1$, the leading first and second order terms are sufficient for the description of the evolution of $f(u,t)$:

$$\frac{\partial f(u,t)}{\partial t} = \frac{\partial}{\partial u}\left[-\frac{(k_1-k_2)N}{\Omega}u(1-u) + \frac{1}{2}\frac{k_1+k_2}{\Omega}\frac{\partial}{\partial u}u(1-u)\right] f(u,t) \tag{40}$$

with boundary conditions $f(0,t) = 0 = f(1,t)$.

This is the Fokker-Planck equation, which is associated to the Itô stochastic differential equation for $U(t)$:

$$dU(t) = \gamma U(1-U)dt + \sigma\sqrt{U(1-U)}dW(t) \tag{41}$$

when we recognize $\gamma = (k_1 - k_2)N/\Omega$ and $\sigma^2 = (k_1 + k_2)/\Omega$, where $W(t)$ is a Wiener Process (usual Brownian motion). In the limit $N \to \infty$ and $\Omega \to \infty$ with $N/\Omega = O(1)$, the coefficient γ is finite and the noise strength $\sigma \sim 1/\sqrt{N}$ goes to zero recovering the deterministic dynamics. Note that $U(t)$ in the stochastic equation is a continuum variable but the discreteness is present in the noise term, giving rise to fluctuations to the system.

For example it gives the possibility for extinction which is not possible in the logistic equation without noise; the addition of the noise indeed allows the process to eventually become extinct (in finite time, no less) due to fluctuations. Indeed, starting from any initial conditions $0 \leq U(0) = u_0 < 1$, the deterministic model predicts $U(t)$ drifts monotonically toward 1 although it never reaches 1 in finite time. The stochastic process, on the other hand, hits $U = 0$ (at a random, almost surely finite, time) with probability p_{ext} which is a function of the initial conditions:

$$p_{ext} = \text{Prob}\left\{U(t) \xrightarrow{t\to\infty} 0 | U(0) = u_0 \geq 0\right\} = \frac{e^{\frac{2\gamma}{\sigma^2}(1-u_0)} - 1}{e^{\frac{2\gamma}{\sigma^2}} - 1}. \tag{42}$$

This is positive for all $0 \leq u_0 < 1$ as shown in Figure 1.

To see this consider the adjoint operator

$$\mathcal{L}^* = u(1-u)\left[\gamma\frac{\partial}{\partial u} + \frac{\sigma^2}{2}\frac{\partial^2}{\partial u^2}\right]$$

and note that the function $h(u)$ that solves $\mathcal{L}^* h(u) = 0$ with boundary conditions $h(0) = 1$ and $h(1) = 0$ is precisely

$$h(u) = \frac{e^{\frac{2\gamma}{\sigma^2}(1-u)} - 1}{e^{\frac{2\gamma}{\sigma^2}} - 1}. \tag{43}$$

Then consider Itô's formula for the increments of the stochastic process $h(U(t))$:

$$\begin{aligned} dh(U(t)) &= h'(U(t))dU + \frac{1}{2}h''(U(t))dU^2 \\ &\approx \left[h'(U(t))\gamma + \frac{\sigma^2}{2}h''(U)\right] U(1-U)dt \\ &\quad + h'(U(t))\sigma\sqrt{U(1-U)}dW \end{aligned} \tag{44}$$

using (41), remembering that $dW^2 \sim dt$, and neglecting small terms proportional to dt^2 and $dtdW$. The first term on the right hand expression is identically 0 because of $\mathcal{L}^* h(u) = 0$. Thus only the term proportional to the increment of the Brownian motion dW remains so $h(U(t))$ is a Martingale, i.e., the expectation of $h(U(t))$ does not change in time:

$$\frac{d}{dt}\mathbb{E}\{h(U(t))\} = 0. \tag{45}$$

Now on the one hand as $t \to \infty$ the process $U(t)$ either goes extinct ($U(t) \to 0$ and $h(U(t)) \to 1$) with probability p_{ext} or it saturates ($U(t) \to 1$ and $h(U(t)) \to 0$) with probability $1 - p_{ext}$. This means that $\mathbb{E}\{h(U(t))\} \to 1 \times p_{ext} + 0 \times (1 - p_{ext}) = p_{ext}$. But the $\mathbb{E}\{h(U(t))\}$ does not change with time so it is equal to its initial value and we conclude that when $U(0) = u_0$,

$$p_{ext} = \mathbb{E}\{h(U(t))\} = \mathbb{E}\{h(U(0))\} = h(u_0) = \frac{e^{\frac{2\gamma}{\sigma^2}(1-u_0)} - 1}{e^{\frac{2\gamma}{\sigma^2}} - 1}. \tag{46}$$

All this naturally leads to study the stochastic version of the FKPP equation obtained from the deterministic equation by the addition of a multiplicative noise like that in (41):

$$\frac{\partial U(x,t)}{\partial t} = D\frac{\partial^2 U}{\partial x^2} + \gamma U(1-U) + \epsilon\sqrt{U(1-U)}\eta(x,t) \tag{47}$$

where ϵ is the strength of the noise and $\eta(x,t)$ is a Guassian white noise, δ-correlated in time and space:

$$\langle \eta(x,t)\eta(y,s)\rangle = \delta(x-y)\delta(t-s).$$

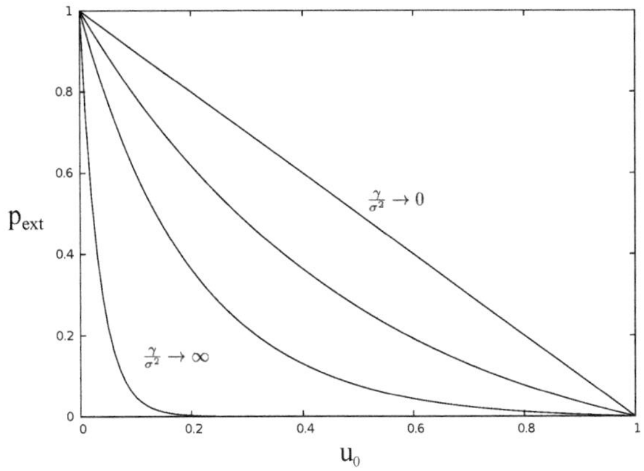

Figure 1. The extinction probability as function of initial condition u_0 for changing the parameter $\frac{\gamma}{\sigma^2}$.

What sort of solutions exist for the sFKPP equation? A theorem assures that for $\epsilon > 0$ solutions enjoy a "compact support property". This means that if the initial data $U(x, 0)$ is compactly supported in space then $U(x, t)$ is also compactly supported for all $t > 0$. Moreover if $U(x, 0)$ vanishes identically for, say, positive x, then $U(x, t)$ vanishes identically for sufficiently large x at all future time $t > 0$. And starting with $U(x, 0) = 1$ for $x \leq 0$ and $U(x, 0) \equiv 0$ for x sufficiently large there is a well defined stochastic front traveling at (almost sure constant) mean speed

$$\bar{c} = \lim_{t \to \infty} t^{-1} \int_0^\infty U(x, t)\, dx$$

that depends on D, γ, and the noise strength ϵ (Mueller and Sowers, 1995).

The primary effect of noise is to extinguish the "leading tail" of the front, and some studies of front propagation modified the growth function in the deterministic FKPP equation in order to take care of discreteness (Brunet and Derrida, 1997, Kessler et al., 1998, and Pechenik and Levine, 1999). The equation is then given by a cutoff in the growth term replacing the classic logistic growth by

$$\gamma u(1 - u)\Theta(u - \frac{1}{N})$$

where the Heaviside step function $\Theta(u - \frac{1}{N})$ is 0 for $u < \epsilon^2$ and 1 for $u \geq \epsilon^2$. This models the level of discreteness $1/N$ according to $\epsilon \sim 1/\sqrt{N}$.

If we look at the mechanical analog this has the same effect of modifying the potential $V(F)$ as shown in Figure 2. If the friction coefficient (the

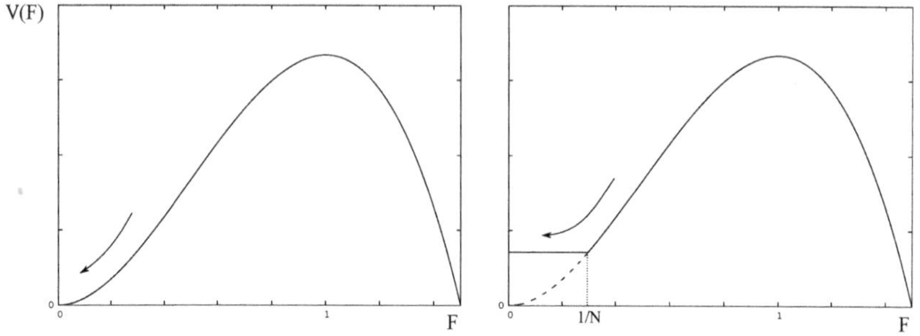

Figure 2. On the right the schematic view of the potential $V(F)$ in the mechanical analog of the differential equation (37). On the left, the potential $V(F)$ modified by the cutoff in order to model the noise in the deterministic FKPP equation.

front speed) c is too small, the solution overshoots zero which is unphysical. On the other hand if c is too large the "particle" never reaches 0, but rather comes to rest between 0 and $1/N$. The conclusion is that there is now a *unique* friction coefficient (front speed) c in order to have a steadily propagating wave front. Note that this unique speed c must be less than the deterministic continuum minimum value $c_{min} = 2\sqrt{D\gamma}$. In fact the mechanical analog implies c is less than the deterministic continuum minimum value c_{min} by corrections of order $(\ln \epsilon^{-2})^{-2}$:

$$c \sim \sqrt{D\gamma}\left[2 - \pi^2 \left(\ln \frac{(D\gamma)^{\frac{1}{2}}}{\epsilon^2}\right)^{-2}\right]. \tag{48}$$

The speed of the front c in the presence of noise is slower than the minimum value for the speed in the deterministic case and as $\epsilon \to 0$ the correction vanishes only very slowly. The expression (48) was recently proved rigorously for the sFKPP by Mueller et al. (2011).

Duality without spatial dependence

In order to analyze the dependence of the front speed in the strong noise limit it is useful to use the duality between the sFKPP equation and the diffusion-reaction process $A+A$. However a preliminary approach to duality without any spatial dependence is convenient. In this section we develop the concept of duality between the solution of the stochastic (Itô) differential equation

$$dU = \gamma U(1-U)dt + \sigma\sqrt{U(1-U)}dW, \tag{49}$$

where dW is the usual Brownian motion, and the well-stirred chemical reaction processes

$$A \xrightarrow{\gamma} A + A$$
$$A + A \xrightarrow{\sigma^2} A.$$

The microscopic Markov Process $N_A(t)$ is described by $P_n(t) = \text{Prob}\{N(t) = n\}$, $n = 1, 2, 3, \ldots$, which evolves according to the master equation

$$\begin{aligned} \frac{d}{dt}P_n(t) &= \gamma(n-1)P_{n-1} - \gamma n P_n - \sigma^2\frac{n(n-1)}{2}P_n + \sigma^2\frac{n(n+1)}{2}P_{n+1} \\ &= \sum_{m=1}^{\infty} M_{nm}P_m(t) \end{aligned} \tag{50}$$

defining the transition matrix

$$M_{nm} = \gamma m\delta_{n,m+1} - \gamma m\delta_{n,m} - \frac{\sigma^2}{2}m(m-1)\delta_{n,m} + \frac{\sigma^2}{2}\delta_{n,m-1}. \tag{51}$$

On other hand a change in variable $Z(t) = 1 - U(t)$ in Eq.(49) gives

$$dZ = -\gamma Z(1-Z)dt + \sigma\sqrt{Z(1-Z)}dW. \tag{52}$$

The process $Z(t)$ has stable fixed point 0 and unstable fixed point 1. The Itô formula applied to the monomial $Z(t)^m$ with integer values $m \geq 1$ yields

$$dZ^m = mZ^{m-1}dZ + \frac{1}{2}m(m-1)Z^{m-2}(dZ)^2. \tag{53}$$

Substitute the equation for dZ into the expression above, keeping only terms of order dW and dt and recalling $(dW)^2 = dt$, we obtain

$$\begin{aligned} dZ^m = &\left(-m\gamma Z^m + m\gamma Z^{m+1} + \frac{\sigma^2}{2}m(m-1)Z^{m-1} - \frac{\sigma^2}{2}m(m-1)Z^m\right)dt \\ &+ \sigma m Z^{m-1}\sqrt{Z(1-Z)}dW \end{aligned} \tag{54}$$

which is

$$dZ^m = \sum_{n=1}^{\infty} Z(t)^n M_{nm}dt + \sigma m Z^{m-1}\sqrt{Z(1-Z)}dW \tag{55}$$

where the matrix M_{nm} turns out to be *exactly* the matrix in (50) for the evolution of P_n. The term proportional to dW vanishes when one takes the expectation, so the conclusion is that moments of $Z(t)$ evolve in time according to

$$\frac{d}{dt}\langle Z(t)^m\rangle = \sum_{n=1}^{\infty}\langle Z(t)^n\rangle M_{nm} \tag{56}$$

where we adopt the notation $\langle\cdot\rangle$ for the expectation $\mathbb{E}\{\cdot\}$.

Choose $T>0$ and consider times $0\le t\le T$. The random variable

$$\mathcal{M}(t) = \sum_{m=1}^{\infty} Z(t)^m P_m(T-t) \tag{57}$$

is well defined since $0\le Z\le 1$ and the $P_m\ge 0$ are summable. And, as will be seen, it is a Martingale (that is, its expectation value does not change with time):

$$\langle\mathcal{M}(t)\rangle = \langle\mathcal{M}(0)\rangle = \langle\mathcal{M}(T)\rangle\,. \tag{58}$$

Proof.

$$\frac{d\langle\mathcal{M}(t)\rangle}{dt} = \sum_{m=1}^{\infty}\left(\frac{d}{dt}\langle Z(t)^m\rangle\right)P_m(T-t) + \sum_{m=1}^{\infty}\langle Z(t)^m\rangle\left(\frac{d}{dt}P_m(T-t)\right)$$

$$= \sum_{m=1}^{\infty}\sum_{n=1}^{\infty}\langle Z(t)^n\rangle M_{nm}P_m(T-t) - \sum_{m=1}^{\infty}\langle Z(t)^m\rangle\sum_{n=1}^{\infty}M_{mn}P_n(T-t) = 0.$$

□

The duality connection follows from the Martigale property. Extending the expectation notation $\langle\cdot\rangle$ to include expectation value over the integer-valued stochastic process $N(t)$ distributed according to $P_n(t)$ we may write

$$\left\langle Z(0)^{N(T)}\right\rangle = \sum_{n=1}^{\infty}\langle Z(0)^n\rangle P_n(T) \tag{59}$$

and

$$\left\langle Z(T)^{N(0)}\right\rangle = \sum_{n=1}^{\infty}\langle Z(T)^n\rangle P_n(0). \tag{60}$$

Thus, according to (58),

$$\left\langle Z(0)^{N(T)}\right\rangle = \left\langle Z(T)^{N(0)}\right\rangle. \tag{61}$$

Returning to the original logistic variable $U(t)$,

$$\left\langle (1-U(t))^{N(0)} \right\rangle = \left\langle (1-U(0))^{N(t)} \right\rangle \tag{62}$$

where the expectation is over the underlying Brownian motion, the stochasticity of the interacting particle process, and over initial conditions.

Using duality is it possible to infer one of the two processes in terms of its initial conditions and the initial and final state of the other process, i.e., it allows the direct computation of the extinction probability p_{ext} for $U(t)$ as explained below. The particle process $N(t)$ has a unique invariant equilibrium (Poisson) distribution

$$P_n^{eq} = \frac{g^n}{n!} \frac{1}{e^g - 1} \quad \text{for} \quad n = 1, 2, \ldots \tag{63}$$

where $g \equiv 2\gamma/\sigma^2$. Duality leaves freedom to choose initial conditions, so let us take the initial condition of the particle process to be the equilibrium distribution itself, $P_n(0) = P_n(T) = P_n^{eq}$. Hence

$$\left\langle (1-U(t))^{N(0)} \right\rangle = \left\langle \sum_{n=1}^{\infty} (1-U(t))^n P_n^{eq} \right\rangle = \left\langle \frac{e^{g(1-U(t))} - 1}{e^g - 1} \right\rangle. \tag{64}$$

On the other hand

$$\left\langle (1-U(0))^{N(T)} \right\rangle = \left\langle \sum_{n=1}^{\infty} (1-U(0))^n P_n^{eq} \right\rangle = \left\langle \frac{e^{g(1-U(0))} - 1}{e^g - 1} \right\rangle. \tag{65}$$

Define the random variable $V(T)$

$$V(T) = \frac{e^{g(1-U(T))} - 1}{e^g - 1} \tag{66}$$

which in the limit $T \to \infty$ has the following values

$$\lim_{T\to\infty} V(T) = \lim_{T\to\infty} \frac{e^{g(1-U(T))} - 1}{e^g - 1} = \begin{cases} 1 & \text{with probability} \quad p_{ext} \\ 0 & \text{with probability} \quad 1 - p_{ext}. \end{cases} \tag{67}$$

So $V(T)$ is an indicator of extinction or fixation and its expectation converges to the probability p_{ext}. Thanks to duality, which gives the equality of expression (64) and (65)

$$p_{ext} = \langle V(T) \rangle = \left\langle \frac{e^{g(1-U(0))} - 1}{e^g - 1} \right\rangle \tag{68}$$

which is the same result of expression (46) when $U(0)$ is the non-random initial condition u_0.

Duality for the sFKPP equation

Now let us include space dependence and develop the duality relation between the sFKPP partial differential equation (33) and the diffusion-reaction particle process. For convenience we consider discretized space, a one-dimensional lattice with spacing h.

The birth-coagulation diffusion process is defined by the local (on-site) reactions

$$A \to A + A \quad \text{at rate} \quad \gamma$$

$$A + A \to A \quad \text{at rate} \quad \sigma^2$$

and the rate of hopping between neighbourhood sites, D/h^2. The system is the Markov process $\mathbf{N}(t) = (\dots, N_{i-1}(t), N_i(t), N_{i+1}(t), \dots)$ specifying the number of particles at each site i at time t. Here we introduce the notation $\mathbf{n} = (\dots, n_{i-1}, n_i, n_{i+1}, \dots)$ where each $0 \leq n_j < \infty$ and the unit vectors $\mathbf{e}_i = (\dots, 0, 1, 0, \dots)$. The probability $P_\mathbf{n} = Prob[\mathbf{N}(t) = \mathbf{n}]$ evolves according to the master equation

$$\frac{d}{dt} P(t)_\mathbf{n} = \sum_{\{\mathbf{m}\}} M_{\mathbf{nm}} P_\mathbf{m}(t) \tag{69}$$

with transition matrix

$$\begin{aligned} M_{\mathbf{nm}} = \sum_i \Bigg[& \gamma m_i \delta_{\mathbf{n},\mathbf{m}+\mathbf{e}_i} - \gamma m_i \delta_{\mathbf{nm}} - \frac{\sigma^2}{2} m_i(m_i - 1)\delta_{\mathbf{nm}} \\ & + \frac{\sigma^2}{2} m_i(m_i - 1)\delta_{\mathbf{n},\mathbf{m}-\mathbf{e}_i} + \frac{D}{h^2} m_{i-1}\delta_{\mathbf{n},\mathbf{m}+\mathbf{e}_i-\mathbf{e}_{i-1}} \\ & - \frac{2D}{h^2} m_i \delta_{\mathbf{nm}} + \frac{D}{h^2} m_{i+1}\delta_{\mathbf{n},\mathbf{m}+\mathbf{e}_i-\mathbf{e}_{i+1}} \Bigg]. \end{aligned} \tag{70}$$

The discrete version of sFKPP equation is

$$dU_i(t) = \left[D\left(\frac{U_{i+1} - 2U_i + U_{i-1}}{h^2} \right) + \gamma U_i(1 - U_i) \right] dt + \sigma\sqrt{U_i(1 - U_i)}dW_i \tag{71}$$

where $W_i(t)$ are independent Brownian motion at each site,

$$dW_i(t)dW_j(t) = \delta_{ij}\, dt.$$

As same as before the change in variable is considered

$$Z_i(t) = 1 - U_i(t)$$

For a given vector of integers $\mathbf{m} = (\ldots, m_{i-1}, m_i, m_{i+1}, \ldots)$, an the application of Itô formula yields

$$d\left(\prod_j Z_j(t)^{m_j}\right) = \sum_{\{\mathbf{n}\}} \left\langle \prod_i Z_i(t)^{n_i} \right\rangle M_{\mathbf{nm}} dt + \sum_i (\text{terms} \propto dW_i)\,. \quad (72)$$

The expectation of the last term vanishes so

$$\frac{d}{dt}\left\langle \prod_j Z_j(t)^{m_j} \right\rangle = \sum_{\{\mathbf{n}\}} \left\langle \prod_i Z_i(t)^{n_i} \right\rangle M_{\mathbf{nm}} \quad (73)$$

where $M_{\mathbf{nm}}$ is exactly the matrix in (70).

For time $0 \leq t \leq T$ it is possible to define the random variable

$$\mathcal{M}(t) = \sum_{\{\mathbf{n}\}} \left(\prod_i Z_i(t)^{n_i}\right) P_{\mathbf{n}}(T-t) \quad (74)$$

which is a Martigale (proved by direct computation as before):

$$\langle \mathcal{M}(t) \rangle = \langle \mathcal{M}(0) \rangle = \langle \mathcal{M}(T) \rangle\,. \quad (75)$$

With analogous passages as before, the duality relationship that holds for any time $t \geq 0$ is

$$\left\langle \prod_i (1 - U_i(0))^{N_i(t)} \right\rangle = \left\langle \prod_i (1 - U_i(t))^{N_i(0)} \right\rangle \quad (76)$$

where the expectation is over the Brownian motion $W_i(t)$, the realization of the particle process $N(t)$, and over initial conditions.

As an application of duality we analyze the long time behavior for the solution of the sFKPP equation with non-random initial condition $U_i(0) = u_i^0$. As $t \to \infty$ all variables $U_i(t)$ become either 0 or they all saturate to 1. Starting from any non-vanishing initial condition the particle process $\mathbf{N}(t)$ goes to equilibrium distribution

$$P_{\mathbf{n}}^{eq} = \prod_i \frac{g^{n_i}}{n_i!} e^{-g}. \quad (77)$$

On one hand

$$\left\langle \prod_i (1-U_i(0))^{N_i(t)} \right\rangle = \sum_{\{\mathbf{n}\}} \prod_i \left(1-u_i^0\right)^{n_i} P_{\mathbf{n}}^{eq}$$
$$= \prod_i \sum_{n=0}^{\infty} \frac{(1-u_i^0)^n g^n}{n!} e^{-g}$$
$$= \prod_i e^{-g u_i^0} = \exp\left\{ -\sum_i \frac{2\gamma}{\sigma^2} u_i^0 \right\}. \quad (78)$$

On the other hand, choosing for $\mathbf{N}(0)$ the equilibrium distribution,

$$\left\langle \prod_i (1-U_i(t))^{N_i(0)} \right\rangle = \left\langle \sum_{\{\mathbf{n}\}} \prod_i (1-U_i(t))^{n_i} P_{\mathbf{n}}^{eq} \right\rangle$$
$$= \left\langle \prod_i e^{g U_i(T)} \right\rangle$$
$$= \left\langle \exp\left\{ -\sum_i \frac{2\gamma}{\sigma^2} U_i(T) \right\} \right\rangle. \quad (79)$$

The random variable in the righthand expression above converges as $t \to \infty$ to the indicator that the process become extinct or not, so its expectation value converges to the probability of extinction p_{ext}. The duality relation thus allows us to infer the exact expression for p_{ext}:

$$p_{ext} = \exp\left\{ -\sum_i \frac{2\gamma}{\sigma^2} u_i^0 \right\}. \quad (80)$$

In the continuum limit $h \to 0$, equation (71) becomes the continuum version of the sFKPP equation

$$\partial_t U(x,t) = D\partial_{xx} U + \gamma U(1-U) + \epsilon\sqrt{U(1-U)}\eta(x,t) \quad (81)$$

with noise strength

$$\epsilon = \sigma\sqrt{h}. \quad (82)$$

The extinction probability is then

$$p_{ext} = \exp\left\{ -\frac{2\gamma}{\epsilon^2} \int_{-\infty}^{\infty} u_0(x)dx \right\} \quad (83)$$

which is exponentially small for large initial mass.

Duality brings another result, too. It allows us to say that the wavefront speed for $U(t)$ is the same for the particle process. To see this, choose the following initial conditions for the two independent processes: only one particle in site $i > 0$ and no particles anywhere else ($\mathbf{N}(0) = \mathbf{e}_i \Leftrightarrow N_j(0) = \delta_{ij}$), and the step function for $U_j(0) = 0$ for any site $j \geq 0$ and $U_j(0) = 1$ for any site $j \leq 0$). Note first that

$$1 - \left\langle \prod_j (1 - U_j(0))^{N_j(t)} \right\rangle = \langle U_i(t) \rangle \tag{84}$$

since the factors on the products are everywhere 1 except at site i. As long as there are no particles on $j \leq 0$ the product above is equal to one, but if there is a particle at any site $j \leq 0$ the product become 0, so it is the indicator function of the fact that particles have arrived in the region $j \leq 0$. The expectation value of this quantity is hence the probability for these events for given particular initial conditions:

$$\langle U_i(t) \rangle = \text{Prob}\left\{\text{any site } j \leq 0 \text{ has a particle at } t | N_j(0) = \delta_{ij}\right\} \tag{85}$$

and we conclude that the wave front for $U_j(t)$ moves with the same speed of the front of $A + A \leftrightarrows A$.

This result is particularly useful in the strong noise limit when $\sigma \to \infty$ or $\epsilon \to \infty$ since, in this case, the s-FKPP equation is dual to the diffusion-controlled particle reaction. The exact speed for this process is known: $v = D\rho_{eq}$ where ρ_{eq} is the density of particles at equilibrium,

$$\rho_{eq} = \frac{2\gamma}{h\sigma^2} = \frac{2\gamma}{\epsilon^2}. \tag{86}$$

Given the equality of the speeds of the two fronts due to duality, this gives

$$c = v = \frac{2D\gamma}{\epsilon^2}. \tag{87}$$

This is the expression for the speed of the front of the sFKPP solutions in the limit of strong noise and it is in excellent agreement with direct numerical simulations; see Doering et al. (2003). (See Doering et al. (2005b) for a discussion of some of the simulation techniques.)

In conclusion the addition of noise in the deterministic FKPP equation is a natural way to consider the effects of discreteness and fluctuations in spatially extended logistic dynamics. The presence of this term can have profound effects on the qualitative behavior on the system. In particular the front speed with noise turns out to be unique, and slower that the minimum value accessible for any deterministic front.

Demographic stochasticity in evolutionary population dynamics

In this last lecture we focus on stochasticity in demographic systems analyzing its effect on the dynamics of competitive species. For the model studied here, the deterministic description produces "degenerate" steady state coexistence of the species with a population ratio determined by the initial conditions. Demographic fluctuations, i.e., birth-death noise, break this degeneracy allowing for the selection of one species over another on an evolutionary time scale that is longer than the deterministic relaxation time but much shorter than the ultimate extinction time. In order to show this behavior we study the system of two competitive species that have the same carrying capacity but different birth and death rates using a Markovian model and compare results with the degenerate deterministic limit. This discussion comes from "Features of Fast Living: On the Weak Selection for Longevity in Degenerate Birth-Death Processes" by Lin, Kim, and the lecturer (Lin et al., 2012).

Deterministic and stochastic models

When a single species grows with constant net growth rate γ, the increment of the population $X(t)$ is proportional to the population itself so

$$\frac{dX(t)}{dt} = \gamma X \tag{88}$$

and the number of individuals grows exponentially in time. If there is a carrying capacity K limiting the total number of individuals, the simple logistic equation

$$\frac{dX}{dt} = \gamma X \left(1 - \frac{X}{K}\right) \tag{89}$$

captures the essential dynamics: $X(t)$ relaxes to the steady state K, achieving $K \pm \mathcal{O}(1)$ in time $t \sim \mathcal{O}(\log K)$ as $K \to \infty$. In natural dimensionless variables $x(t) \equiv \frac{X(\frac{t}{\gamma})}{K}$,

$$\dot{x} = x(1-x) \tag{90}$$

and $x(t)$ reaches $1 \pm 1/K$ in time $t \sim \mathcal{O}(\log K)$ as $K \to \infty$.

The deterministic rate equation does not display finite time extinction but discrete stochastic dynamic models do. Consider the integer-valued continuos time process X_t with birth and death events rates, respectively, β and $\delta(1+n/\tilde{K})$ when $X_t = n$, where $\tilde{K} \equiv K/(\rho - 1)$ and $\rho \equiv \beta/\delta > 1$. The state of the population is then described by the probability $p_n(t) = Prob\{X_t = n\}$ that follows the master equation

$$\begin{aligned} \frac{dp_n(t)}{dt} &= -\beta n p_n + \beta(n-1)p_{n-1} - \delta\left(1 + \frac{n}{\tilde{K}}\right) n p_n \\ &+ \delta\left(1 + \frac{n+1}{\tilde{K}}\right)(n+1)p_{n+1}. \end{aligned} \tag{91}$$

This formulation introduces something interesting into the system: finite time extinction with probability 1 in time $t = \mathcal{O}(e^{cK})$ for large K (Doering et al., 2005a). Before extinction and after the $\mathcal{O}(\log K)$ deterministic relaxation time, population fluctuates $\pm\mathcal{O}(\sqrt{K})$ around the carrying capacity K. The normalised continuous time process

$$x_t = \frac{X_{t/(\beta-\delta)}}{K} \tag{92}$$

has level of discreteness $\delta x = 1/K \ll 1$ and for large K and long time intervals its statistics mimic those of solutions of the Itô stochastic differential equation

$$dx_t = \gamma x_t(1 - x_t)dt + \epsilon\sqrt{\gamma x_t\left(\frac{\rho+1}{\rho-1} + x_t\right)}dW_t \tag{93}$$

where $\gamma = \beta - \delta$ and the noise amplitude $\epsilon = 1/\sqrt{K}$. The multiplicative white noise produces the finite time extinction with probability 1 in time $t = \mathcal{O}(e^{c'K})$ as well (Doering et al., 2005a).

Degenerate model

According to evolutionary dynamics the best-fit species should survive longer than the less-fit. Consider the Markovian birth-death process of two species with t_{ext} of order e^{cK}. We are interested in finding if one of the two species has an extinction time significantly shorter than e^{cK} so that it might be possible to refer the cause of extinction as competition rather than natural inevitable extinction. We consider two species $X(t)$ and $Y(t)$ and assume that one reproduces and also dies fast, whereas the other is slow in reproduction but also lives longer. They compete equally for the same resources and the only difference is on their rates of birth and death.

Start with the simplest description considering the following deterministic equations:

$$\dot{X} = \gamma_X X\left(1 - \frac{X+Y}{K}\right) \quad \text{and} \quad \dot{Y} = \gamma_Y Y\left(1 - \frac{X+Y}{K}\right). \tag{94}$$

Suppose the total number of individuals in the steady state is K and let scale $X(t)$ and $Y(t)$ by normalize them by K, moreover let define $\gamma = \gamma_X/\gamma_Y$ and without loss of generality let consider the case $\gamma > 1$, so that X is the faster species:

$$\dot{x} = \gamma x\,(1 - x - y) \quad \text{and} \quad \dot{y} = y\,(1 - x - y)\,. \tag{95}$$

Trajectories in the phase space satisfy

$$\gamma \frac{dy}{dx} = \frac{y}{x} \tag{96}$$

so that at any instants of time

$$\frac{y(t)}{y(t_0)} = \left(\frac{x(t)}{x(t_0)}\right)^{\frac{1}{\gamma}}\,. \tag{97}$$

Trajectories are shown in Figure 3. From any initial point inside the first quadrant solutions evolve to the stable line of fixed points corresponding to degenerate states of coexistence. This occurs roughly in time of $\mathcal{O}(log K)$, and the final ratio of the two species depends only in the initial conditions. But is deterministic degeneracy real? That is, do the species coexist for an $\mathcal{O}(e^{cK})$ time? If not, which among two species has an advantage and (probably) survives longer?

Demographic stochasticity is taken into account by moving to the Markov processes X_t and Y_t. The random birth events have rates β_X and β_Y and the death events have rates

$$\delta_X\left(1 + \frac{n+m}{\tilde{K}}\right) \qquad \text{and} \qquad \delta_Y\left(1 + \frac{n+m}{\tilde{K}}\right)$$

when $X_t = n$ and $Y_t = m$, where $\tilde{K} = K/(\rho - 1)$ with $\beta_X/\delta_X = \beta_Y/\delta_Y = \rho > 1$. With this formulation $p_{n,m}(t) = \text{Prob}\{X_t = n \text{ and } Y_t = m\}$ evolves according to the master equation

$$\begin{aligned}
\frac{d}{dt}p_{n,m}(t) = &-\left(\beta_X + \delta_X\left[1 + \frac{(n+m)}{\tilde{K}}\right]\right) n p_{n,m} \\
&-\left(\beta_Y + \delta_Y\left[1 + \frac{(n+m)}{\tilde{K}}\right]\right) m p_{n,m} \\
&+\beta_X(n-1)p_{n-1,m} + \delta_X\left[1 + \frac{(n+1+m)}{\tilde{K}}\right](n+1)p_{n+1,m} \\
&+\beta_Y(m-1)p_{n,m-1} + \delta_Y\left[1 + \frac{(n+m+1)}{\tilde{K}}\right](m+1)p_{n,m+1}\,.
\end{aligned} \tag{98}$$

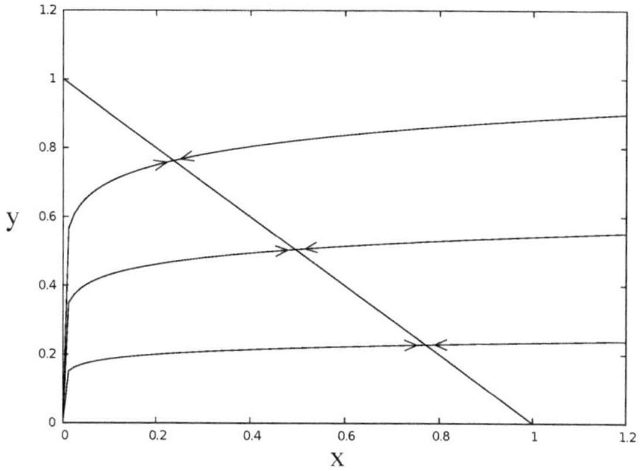

Figure 3. Deterministic trajectories according to equation (97) with $\gamma = 10$ in the (y,x) space.

The coefficients γ_X and γ_Y appearing in the deterministic differential equations (94) are $\gamma_X = \beta_X - \delta_X$ and $\gamma_Y = \beta_Y - \delta_Y$.

In the limit of infinite K the real-valued processes

$$x_t = \frac{X_{t/\gamma_Y}}{K} \qquad \text{and} \qquad y_t = \frac{Y_{t/\gamma_Y}}{K} \tag{99}$$

closely follow (94), but for large but finite K their statistics are well approximated by those of the Markov diffusion processes solving the Itô stochastic differential equations

$$dx_t = \gamma x_t(1 - x_t - y_t)dt + \epsilon\sqrt{\gamma x_t \left(\frac{\rho+1}{\rho-1} + x_t + y_t\right)} dW_t^x \tag{100}$$

$$dy_t = y_t(1 - x_t - y_t)dt + \epsilon\sqrt{y_t \left(\frac{\rho+1}{\rho-1} + x_t + y_t\right)} dW_t^y \tag{101}$$

where dW_t^x and dW_t^y are independent Wiener processes and $\gamma = \delta_X/\delta_Y = \beta_X/\beta_Y$. The noise amplitude $\epsilon = 1/\sqrt{K}$. Note that for the fast species, γ affects both the relaxation term and the noise term in a way that both terms are larger for the faster species.

For large K the system at initial state (x_0, y_0) follows the deterministic dynamics of (97) to a neighborhood of the degenerate line in time $\mathcal{O}(logK)$, and then it is affected by fluctuations of amplitude of $\mathcal{O}(1/\sqrt{K})$ around it. To decide the victor of the competition we study the reduced process z_t defined as the difference of the two processes x_t and y_t:

$$z_t = x_t - y_t \in [-1, 1]. \tag{102}$$

This represents drift and fluctuation on the coexistence line and it is useful for analyzing the probability of eventual absorptions of z_t at the borders 1 or -1, which means the extinction respectively of y or x. Starting form z_0, the process in z_t may be approximated by a Markovian drift-diffusion process described by drift $v(z)$ and diffusion $D(z)$:

$$dz_t = v(z_t)dt + \sqrt{2D(z_t)}dW_t. \tag{103}$$

Let τ_z be the random time at which z_t hits ± 1 starting from $z \in (-1, 1)$. In order to study the probability of domination of one species over the other we need to know $u(z) \equiv \text{Prob}\{z(\tau_z) = -1|z_0 = z\}$. The mean time to extinction $m(z) \equiv E\{\tau(z)\}$ sets the time scale for the decision. To compute these we need to determine the functions $v(z)$ and $D(z)$, i.e., to quantify the drift and diffusion along the coexistence line.

Suppose the system is in the position $z_0 = (x_0, y_0)$ in the coexistence line at time $t = 0$. A displacement at time t' due to fluctuations brings the system to the state $(x', y') = (x_0 + \phi a, y_0 + \eta b)$ where ϕ and η are independent random variables which take values ± 1 with probability $1/2$. To be consistent with the stochastic equations (101) a and b should be proportional to the noise and, noting that $x_0 + y_0 \simeq 1$ in the coexistence line,

$$a \sim \epsilon\sqrt{\gamma x_0 \frac{2\rho}{\rho - 1}dt} \qquad \text{and} \qquad b \sim \epsilon\sqrt{y_0 \frac{2\rho}{\rho - 1}dt}. \tag{104}$$

We write

$$a^2 = C\epsilon^2 \gamma x_0 \frac{2\rho}{\rho - 1}dt \quad \text{and} \quad b^2 = C\epsilon^2 y_0 \frac{2\rho}{\rho - 1}dt \tag{105}$$

where $C > 0$ is a constant of order $\mathcal{O}(1)$.

After the random displacement we suppose that the system responds with the deterministic trajectory of Eq.(97) to the coexistence line to position $(x_0 - \xi, y_0 + \xi)$ so that the net displacement in the line is -2ξ, and the drift and diffusion functions are

$$v(z) = \frac{\langle dz_t \rangle}{dt} = -\frac{2\langle \xi \rangle}{dt} \tag{106}$$

and

$$2D = \frac{\langle dz_t^2 \rangle}{dt} = \frac{4\langle \xi^2 \rangle}{dt} \tag{107}$$

where $\langle \cdot \rangle$ means the average over fluctuations. Using $\langle \Phi \rangle = 0 = \langle \eta \rangle$, $\langle \Phi^2 \rangle = 1 = \langle \eta^2 \rangle$ and assuming that ξ can be written as an asymptotic expansion on the noise ϵ, it is possible to write the drift and diffusion as explicit function of z:

$$v(z) = -\frac{2\langle \xi \rangle}{dt} \sim C\epsilon^2 \frac{2\rho\gamma(1-\gamma)}{\rho - 1} \times \frac{1-z^2}{[(1-z)+\gamma(1+z)]^2} \tag{108}$$

$$D(z) = \frac{2\langle \xi^2 \rangle}{dt} \sim C\epsilon^2 \frac{2\rho\gamma}{\rho - 1} \times \frac{1-z^2}{(1-z)+\gamma(1+z)}. \tag{109}$$

Both drift and diffusion are of order $\epsilon^2 \sim 1/K$, so the time scale for selecting between the two species is of order K which is longer than the time for going to the coexistence line, but small compared to ultimate extinction! Moreover, $u(z)$ turns out to be independent of C. Its explicit form, valid for $K \to \infty$, is

$$u(z) = \frac{1-z}{2}\left[1 + \left(\frac{\gamma - 1}{\gamma + 1}\right)\frac{1+z}{2}\right]. \tag{110}$$

As can be seen from Figure 4, the form of $u(z)$ means that the longer lived species is favored, but note that starting with a 50%-50% initial population, $z = 0$, the probability of Y winning is always constrained to be between 25% and 75%.

In a different way, the time for the extinction of one species $m(z)$ depends explicitly on C, which is a constant to be determined. In order to do that, one can consider the truly degenerate state in which the two species are absolutely identical but different only for labelling, so $\gamma = 1$. In this case $m(z)$ satisfies

$$C\epsilon^2\left(\frac{\rho}{\rho - 1}\right)\left(1 - z^2\right)\frac{d^2m}{dz^2} = -1 \tag{111}$$

with boundary conditions $m(\pm 1) = 0$. The comparison between ODE (111) and the leading terms approximation of the mean first passage time of the process (x_t, y_t) to the borders allows to conclude that $C = 1$.

Numerical Monte Carlo simulations of the Markov process defined by the master equation (98) varying γ and K, (while fixing ρ) show that the asymptotic theory is in agreement with the data, although data for the mean extinction time systematically underestimate the asymptotic theory for high values of γ (Lin et al., 2012).

In conclusion demographic fluctuations can influence the selection of one species over another. The problem in this scenario is an example of a technique for solving this kind of problems with the intention of understanding how generalize it to more complex systems where fluctuations are due, for

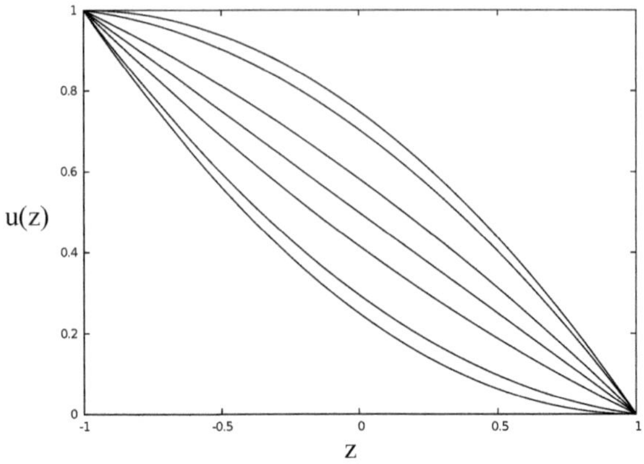

Figure 4. Probability of domination of Y-species over X starting from position z in the coexistence line. From bottom to top $\gamma \to 0$, $\gamma = 0.1, 0.5, 1, 2, 10$ and $\gamma \to \infty$.

example, to births and deaths and spatial dispersion of individuals in homogeneous or heterogeneous environments. Then fluctuations can affect the selection or the extinction of one the species or another (Waddell et al., 2010).

Bibliography

D. ben Avraham. Complete exact solution of a diffusion-limited coalescence, $A + A \to A$. *Physical Review Letters*, 81(21):4756–4759, 1998.

D. ben Avraham and C. R. Doering. Equilibrium of two-species annihilation with input. *Physical Review A*, 37(12):5007–5009, 1988.

D. ben Avraham, M. A. Burschka, and C. R. Doering. Statics and dynamics of a diffusion-limited reaction: Anomalous kinetics, nonequilibrium self-ordering, and a dynamic transition. *Journal of Statistical Physics*, 60: 695–728, 1990.

O. Bénichou, C. Chevalier, J. Klafter, B. Meyer, and R. Voituriez. Geometry-controlled kinetics. *Nature Chemistry*, 2:472, 2010.

M. Bramson. Convergence of solutions of the Kolmogorov equations to traveling waves. *Memoires of the American Mathematical Society*, 44, 1983.

M. Bramson and L. Lebowitz. Asymptotic behavior of densities for two-particle annihilating random walks. *Journal of Statistical Physics*, 62: 297–372, 1991.

E. Brunet and B. Derrida. Shift in the velocity of a front due to a cutoff. *Physical Review E*, 56(3):2597–2604, 1997.

E. Clément, L. M. Sander, and R. Kopelman. Steady-state diffusion-controlled $A + B \to 0$ reactions in two and three dimensions: Rate laws and particle distributions. *Physical Review A*, 39(12):6466–6471, 1989.

C. R. Doering and M. A. Burschka. Long crossover times in a finite system. *Physical Review Letters*, 64(3):245–248, 1990.

C. R. Doering, M. A. Burschka, and W. Horsthemke. Fluctuations and correlations in a diffusion-reaction system: Exact hydrodynamics. *Journal of Statistical Physics*, 65:953–970, 1991.

C. R. Doering, C. Mueller, and P. Smereka. Interacting particles, the stochastic Fisher-Kolmogorov-Petrovsky-Piscounov equation, and duality. *Physica A*, 325:243–259, 2003.

C. R. Doering, K. V. Sargsyan, and L. M. Sander. Extinction times for birth-death processes. *SIAM Journal on Multiscale Modeling and Simulation*, 3:283–299, 2005a.

C. R. Doering, K. V. Sargsyan, and P. Smereka. A numerical method for some stochastic differential equations with multiplicative noise. *Physics Letters A*, 344:149–155, 2005b.

U. Ebert and W. van Saarloos. Front propagation into unstable states: universal algebraic convergence toward uniformly translating pulled fronts. *Physica D*, 146:1–99, 2000.

D. A. Kessler, Z. Ner, and L. M. Sander. Front propagation: Precursors, cutoffs, and structural stability. *Physical Review E*, 58(1):107–114, 1998.

R. Kopelman. Fractal reaction kinetics. *Science*, 241:1620–1626, 1988.

R. Kroon, H. Fleurent, and R. Sprik. Diffusion-limited exciton fusion reaction in one-dimensional tetramethylammonium manganese trichloride (TMMC). *Physical Review E*, 47:2462–2472, 1993.

Y. T. Lin, H. Kim, and C. R. Doering. Features of fast living: On the weak selection for longevity in degenerate birth-death progresses. *Journal of Statistical Physics*, 148:646–662, 2012.

E. Monson and Raoul Kopelman. Observation of laser speckle effects and nonclassical kinetics in an elementary chemical reaction. *Physical Review Letters*, 85(3):666–669, 2000.

C. Mueller and R. B. Sowers. Random traveling waves for the KPP equation with noise. *Journal of Functional Analysis*, 128, 1995.

C. Mueller, L. Mytnik, and J. Quastel. Effect of noise on front propagation in reaction-diffusion equations of KPP type. *Inventiones Mathematicae*, 184:405–453, 2011.

A. A. Ovchinnikov and Ya. B. Zeldovich. Role of density fluctuations in bimolecular reaction kinetics. *Chemical Physics*, 28:215–218, 1978.

L. Pechenik and H. Levine. Interfacial velocity corrections due to multiplicative noise. *Physical Review E*, 59(4):3893–3900, 1999.

D. Toussaint and F. Wilczek. Particle-antiparticle annihilation in diffusive motion. *J. Chem. Phys.*, 78:2642–2647, 1983.

J. N. Waddell, L. M. Sander, and C. R. Doering. Demographic stochasticity versus spatial variation in the competition between fast and slow dispersers. *Theoretical Population Biology*, 77:279–286, 2010.

Multiscale Crowd Dynamics Modeling and Theory

Andrea Tosin*

* Istituto per le Applicazioni del Calcolo "M. Picone"
Consiglio Nazionale delle Ricerche, Rome, Italy

Abstract This chapter deals with models of living complex systems, chiefly human crowds, by methods of conservation laws and measure theory. We introduce a modeling framework which enables one to address both discrete and continuous dynamical systems in a unified manner using common phenomenological ideas and mathematical tools as well as to couple these two descriptions in a multiscale perspective. Furthermore, we present a basic theory of well-posedness and numerical approximation of initial-value problems and we discuss its implications on mathematical modeling.

1 Introduction

By *living complex systems* we mean multi-agent systems composed by living entities, which take part in *group dynamics* while trying to chase *individual purposes.* Specifically, in this chapter we will be concerned with human crowds. We will assimilate pedestrians to *active particles*, the *activity* being their ability to set one or more intermediate and final goals (such as e.g., avoiding collisions with other particles, reaching a destination) and to act directly on their own dynamics to chase them, without being passively prone to external influences. This gives rise to *collective* dynamics based primarily on *individual* behavioral rules. At times active particles cooperate for chasing a group goal, like in *consensus* and *rendez-vous* problems studied by Canuto et al. (2012). On other occasions, instead, they do not cooperate consciously, which makes group dynamics more difficult to be predicted and nevertheless often surprisingly ordered and coordinated: it is the so-called *self-organization*, see Cristiani et al. (2010).

In order to model such systems it is necessary to set up mathematical structures suitable to cope with their complexity, partly due to that overall dynamics are ultimately *multiscale.* In fact, they originate from individual behaviors at the *microscopic* scale of single particles. Next, they are "amplified" by interactions among particles up to producing collective behaviors at the *macroscopic* scale of the group, which cannot be directly deduced from the knowledge of individual ones. Moreover, the collective state of the group can in turn impact locally on the behavioral rules adopted by single particles.

A large class of mathematical models uses *conservation* (or *balance*) *laws*, expressing the fact that some physical quantities, such as mass, linear momentum, and energy of the system, either do not change during the evolution or change in consequence of specific production/destruction mechanisms. However, living systems can hardly be confined in strict balance principles. For instance, their ability to elaborate *behavioral strategies* for chasing a purpose makes them continuously put and remove energy from the system in unconventional manners. Indeed entropy principles classically related to the equiprobability of the states may be questioned, for self-organization promotes special, usually inhomogeneous, configurations to the detriment of more generic and homogeneous ones as stated by Schröedinger (1967). Moreover, it may be difficult to ascribe the variations of linear momentum to possibly "generalized" forces, because the dynamics of living systems are not ruled purely by inertia. Of course, active particles do not elude usual physical laws, rather they can influence them by means of personal decisions, whose effects are not necessarily describable in terms of external force fields. In other words, a straightforward application of the very same ideas successfully used to describe other kinds of passive systems may not completely fit active particles, because this analogy would forcedly miss some distinctive features that heavily differentiate living from inert matter.

Among all classical balance principles mentioned above, probably the less questionable one for the systems at hand is the conservation of mass: when describing the evolution in space and time of human crowds it makes sense to assume that no proliferation or destruction of pedestrians occur. Notice that this does not imply by itself any specific dynamics of the interactions among the individuals, it simply requires the conservation of their number. Thus interactions can still be mechanical or non-mechanical, local or nonlocal, binary or multiple, and so on.

Starting from the mass conservation principle, in this chapter we describe a unified mathematical framework which allows one to model crowd dynamics by embedding the discrete description of individual pedestrians and the continuous one of the collectivity. The key point is the reinterpretation of the continuity equation in terms of abstract mass measures featuring a singular component (Dirac deltas), which represents the discrete level, and an absolutely continuous one (with respect to the Lebesgue measure), which represents the continuous level. In more detail, in Section 2 we first introduce the abstract equation and then specialize it to the case of pedestrian interaction models. In Section 3 we discuss the use of the measure formulation for obtaining discrete, continuous, and multiscale models, relating furthermore the structure of the measure and the behavioral strategy of pedestrians. In Section 4 we present the basic qualitative results concerning well-posedness and numerical approximation of the Cauchy problem for the mathematical structures previously deduced. Finally, in Section 5 we discuss the relevance of these qualitative results as guidelines for the construction of specific models which are both physically realistic and mathematically robust.

2 Mathematical models by time-evolving measures

From the mathematical point of view, the *mass* of a d-dimensional system ($d = 1, 2, 3$ for physical purposes) at time t is a Radon positive measure $\mu_t : \mathcal{B}(\mathbb{R}^d) \to \mathbb{R}_+$ defined over the Borel σ-algebra $\mathcal{B}(\mathbb{R}^d)$ in the physical space $\mathbb{R}^d$. In our case, for all measurable set $E \in \mathcal{B}(\mathbb{R}^d)$ the number $\mu_t(E) \geq 0$ gives an estimate of the crowding of the region $E \subseteq \mathbb{R}^d$ at time t (ideally, it can be thought of as the "average" number of pedestrians occupying the region E at time t). In particular, if we consider a crowd composed by N pedestrians then, owing to the mass conservation principle, we must have $\mu_t(\mathbb{R}^d) = N$ for all t. This can be expressed in differential form by saying that the measure μ_t satisfies the equation:

$$\frac{\partial \mu_t}{\partial t} + \nabla \cdot (\mu_t v_t) = 0, \tag{1}$$

where $v_t = v_t(x) : \mathbb{R}^d \to \mathbb{R}^d$ is, at time t, a transport velocity field. Equation (1) is written in a formal fashion but has to be properly understood in the weak sense of distributions. For all test function $\phi \in C_c^\infty(\mathbb{R}^d)$ and for all $t \in (0, T_{\max}]$, $T_{\max} > 0$ being a final time, it means:

$$\int_{\mathbb{R}^d} \phi(x)\, d\mu_t(x) = \int_{\mathbb{R}^d} \phi(x)\, d\mu_0(x) + \int_0^t \int_{\mathbb{R}^d} v_s(x) \cdot \nabla\phi(x)\, d\mu_s(x)\, ds, \tag{2}$$

where μ_0 is a positive Radon measure to be assigned, which represents the initial distribution of the crowd. If the transport velocity is bounded, i.e., there exists a constant $V_{\max} > 0$ such that

$$|v_t(x)| \leq V_{\max}, \qquad \forall x \in \mathbb{R}^d,\ t \in (0, T_{\max}],$$

it is not difficult to show that (2) implies indeed $\mu_t(\mathbb{R}^d) = \mu_0(\mathbb{R}^d)$ for all $t \in (0, T_{\max}]$, hence actually the conservation of the total number of pedestrians

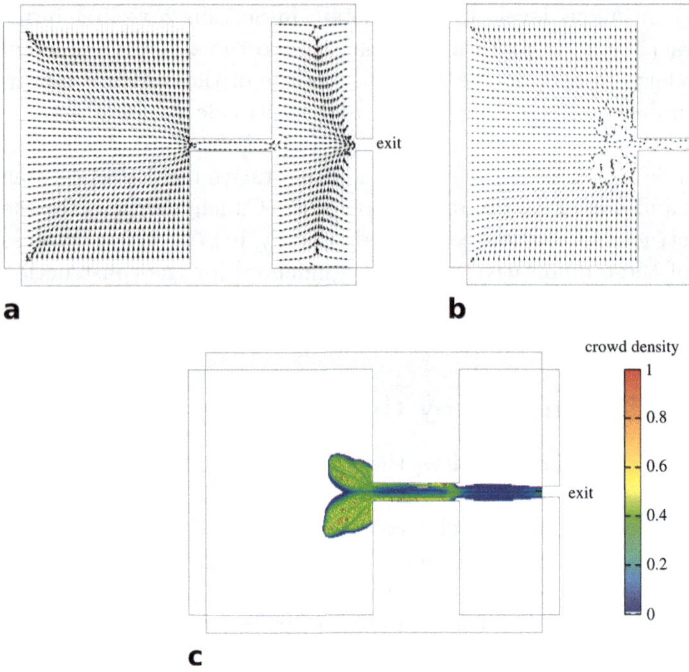

Figure 1. a. Example of a desired velocity field v_{d} in a domain formed by two regions communicating through a narrow corridor. **b.** "True" velocity v (cf. Equation (3)) when the interaction velocity v_{i} among pedestrians is added to the desired velocity. **c.** Crowd density field with generates the true velocity illustrated in **b.**

as fixed at the initial time. It is sufficient to take a sequence $\{\phi_m\}_{m\geq 1} \subseteq C_c^\infty(\mathbb{R}^d)$ of test functions such that $0 \leq \phi_m(x) \leq 1$ for all $x \in \mathbb{R}^d$ and all $m \geq 1$, with in addition $\phi_m(x) \to 1$ and $\nabla\phi_m(x) \to 0$ pointwise for $m \to \infty$, and then invoke the Dominated Convergence Theorem (the existence of such a sequence is guaranteed by Uryshon's Lemma).

2.1 Modeling pedestrian interactions

Equation (1), or alternatively (2), gives the time evolution of the crowd distribution μ_t provided a transport velocity is assigned. Recalling the discussion set forth in the Introduction, given the lack of a balance of linear momentum to be coupled to the mass conservation it is necessary to model directly the field v_t.

Other crowd models available in the literature follow similar ideas, see e.g., Coscia and Canavesio (2008); Colombo and Rosini (2009). Typically, pedestrian velocity is obtained from an empirical constitutive relationship, the so-called *fundamental diagram*, which expresses it as a known function of the distribution in space of the crowd in (locally) stationary homogeneous conditions. Here we propose instead a modeling of v_t more focused on pedestrian interactions, in order to ground the dynamics directly on the idea of active behavior discussed in the Introduction.

We write the velocity as the sum of two contributions:

$$v_t(x) := v_{\text{d}}(x) + v_{\text{i}}[\mu_t](x), \tag{3}$$

where square brackets indicate a functional dependence on the measure μ_t.

The function $v_{\text{d}} : \mathbb{R}^d \to \mathbb{R}^d$ is the *desired velocity*, i.e., the velocity at which an isolated pedestrian would head for her destination. It is independent of the system dynamics, being determined *a priori* only by the geometry of the domain, including the presence of possible obstacles viewed as holes in $\mathbb{R}^d$, namely regions that pedestrian cannot access (Figure 1a). Conversely, the function $v_{\text{i}}[\mu_t] : \mathbb{R}^d \to \mathbb{R}^d$ is the *interaction velocity*, i.e., the correction that pedestrians make to the desired velocity due to mutual interactions. It takes into account that individuals generally aim at avoiding crowded areas, hence it adds a *repulsive* contribution to v_{d} (Figures 1b-c). Moreover, its effect is *nonlocal*, because pedestrians anticipate their own decisions through a process of synthesis of the information about the crowd distribution in the immediate vicinity. Out of these arguments, we set:

$$v_{\text{i}}[\mu_t](x) = \int_{\mathbb{R}^d} K(x, y)\eta_{\mathcal{S}(x)}(y)\, d\mu_t(y), \tag{4}$$

where:

- $K : \mathbb{R}^d \times \mathbb{R}^d \to \mathbb{R}^d$ is the *interaction kernel*, which describes the repulsion acting on the individual in x, called *test pedestrian*, because of the presence of an individual in y, called *field pedestrian*. Generally, recalling also the Galileian invariance, the function K depends on x, y through their distance $|y - x|$ along the segment connecting them. A prototypical example is:

$$K(x, y) \sim -\frac{1}{|y-x|} \cdot \frac{y-x}{|y-x|}$$

  for $|y - x|$ “not too” small, whereas for $y \to x$ it may be necessary to introduce a regularization in order to avoid singularities (cf. Section 5);

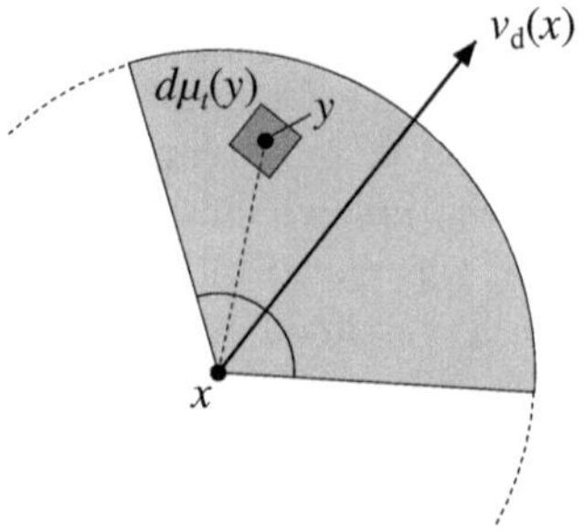

Figure 2. Sensory region of the test pedestrian in x.

- $\eta_{\mathcal{S}(x)} : \mathbb{R}^d \to \mathbb{R}_+$ is a *cut-off function* limiting the influence on the test pedestrian in x to field pedestrians within her *sensory region* $\mathcal{S}(x) \subset \mathbb{R}^d$. Typically, $\eta_{\mathcal{S}(x)}$ is smooth and compactly supported in $\mathcal{S}(x)$ (for instance, it can be a mollified version of the characteristic function of the set $\mathcal{S}(x)$). A prototypical sensory region is a circular sector centered in x, symmetric with respect to the local direction of the desired velocity $v_{\mathrm{d}}(x)$, and oriented along the latter (Figure 2). This models *anisotropic* interactions: the test pedestrian is affected by field pedestrians ahead but not behind. The radius of the circular sector is the maximal distance at which a field pedestrian can have an influence on the test pedestrian, while the angle of the sector identifies the visual cone of the latter.

3 Multiscale approach

Equations (1), (3), (4) provide a unified modeling framework which comprises both *discrete* and *continuous* dynamics. The key point is the *spatial structure* of the measure μ_t.

Discrete dynamics are obtained if the spatial structure of μ_t is discrete:

$$\mu_t = \sum_{i=1}^{N} \delta_{x_i(t)}, \tag{5}$$

$\{x_i(t)\}_{i=1}^N \subset \mathbb{R}^d$ being the set of all and only points where pedestrians are distributed at time t. Plugging (5) into (2), (4) yields:

$$\dot{x}_i = v_{\mathrm{d}}(x_i) + \sum_{j=1}^{N} K(x_i, x_j)\eta_{\mathcal{S}(x_i)}(x_j) \qquad (i = 1, \dots, N), \tag{6}$$

which completely characterizes the evolution of the crowd distribution.

Continuous dynamics are instead obtained if the spatial structure of μ_t is continuous, i.e., if mass and volume are proportional (in the sense of Radon-Nikodym's Theorem):

$$d\mu_t(x) = \rho_t(x)\,dx, \tag{7}$$

where now $\rho_t : \mathbb{R}^d \to \mathbb{R}_+$ is the *mass density* of the crowd distribution at time t, such that $\int_{\mathbb{R}^d} \rho_t(x)\,dx = N$ (thus, in particular, $\rho_t \in L^1(\mathbb{R}^d)$ for all t). In this frame, the support of ρ_t in $\mathbb{R}^d$ is conceptually the counterpart of the set $\{x_i(t)\}_{i=1}^N$ above. Inserting (7) in (2), (4) gives (actually a weak form of):

$$\frac{\partial}{\partial t}\rho_t(x) + \nabla \cdot \left[\rho_t(x)\left(v_\mathrm{d}(x) + \int_{\mathbb{R}^d} K(x,\,y)\eta_{S(x)}(y)\rho_t(y)\,dy\right)\right] = 0, \tag{8}$$

namely a conservation law with nonlocal flux, which in turn characterizes completely the evolution of the distribution of pedestrians.

Do Equations (6), (8) describe the same system and the same dynamics? Yes and no. They formalize two different mathematical models of the same physical system, which however originate from the common abstract structure (1)–(4). Therefore they share the phenomenological description of the *individual microscopic interactions*, expressed by the kernel K and by the cut-off function $\eta_{S(\cdot)}$, but can predict different *collective macroscopic effects* because the latter depend on the spatial structure of the measure μ_t in Equation (4).

More precisely, it is useful to understand the spatial structure of μ_t as the modeling counterpart of the *perception* of the test pedestrian, which affects the way the latter reacts to surrounding individuals. A discrete perception can be typical of sparse crowds or of leisure-type travel purposes, when pedestrians are more sensitive to the *one-by one* distribution of their neighbors. Conversely, a continuous perception can be typical of dense crowds or of business-type travel purposes (e.g., commuters in rush hours), when pedestrians tend to interact with *subgroups* of other walkers as a whole. This corresponds to the expression of a *behavioral strategy*, which can impact in a non-negligible manner on the classical laws of motion (see also Bruno et al. (2011) for a different modeling approach to the concept of perception, however always related to the expression of a behavioral strategy by pedestrians). Therefore, although the elementary pairwise interaction rules are always the same, their collective effect can be greatly different due to the "filtering" operated by perception. Such a phenomenology is possible because crowds are *granular living* systems: the number of pedestrians is locally always finite and "small" (at least compared to the order of magnitude of the Avogadro's number in classical gas dynamics), which makes large-scale dynamics rather sensitive to individual behaviors on smaller scales.

The concept of perception can be formalized in the model by means of a parameter $\theta \in [0, 1]$, which determines a *scale* of spatial structures of the measure μ_t:

$$\mu_t = \theta \sum_{j=1}^{N} \delta_{x_j(t)} + (1-\theta)\rho_t \mathcal{L}^d, \tag{9}$$

$\mathcal{L}^d$ being the Lebesgue measure in $\mathbb{R}^d$. By plugging this representation into Equation (4) we see that the transport velocity (3):

$$\begin{aligned} v_t(x) = v_{\mathrm{d}}(x) + \theta \sum_{j=1}^{N} K(x,\, x_j)\eta_{S(x)}(x_j) \\ + (1-\theta)\int_{\mathbb{R}^d} K(x,\, y)\eta_{S(x)}(y)\rho_t(y)\, dy \end{aligned} \tag{10}$$

depends now on a weighted contribution of discrete and continuous dynamics. Clearly, the two choices discussed above correspond to the particular cases $\theta = 0$ (continuous dynamics) and $\theta = 1$ (discrete dynamics). Nevertheless, if $0 < \theta < 1$ this formalism allows one to deal, more in general, with hybrid dynamics which are neither fully discrete nor fully continuous (cf. the case studies presented in Figures 3, 4). Moreover, the transport of the measure (9) by means of Equation (1) with the velocity field (10) allows also for a purely continuous representation of the crowd distribution evolving according to genuinely discrete dynamics or, conversely, a purely discrete representation evolving according to genuinely continuous dynamics. Finally, if the perception parameter θ changes in space, i.e., it is converted into a function of x, then the model enables one to have different types of dynamics in different sub-domains (see e.g., Cristiani et al. (2012), where this idea is used in the case of vehicular traffic for coupling continuous dynamics along straight roads and discrete dynamics at crossroads, where driver perception is sharpened by vehicles coming from different merging directions).

4 Basic theory

Once an initial measure, say $\bar{\mu}$, is prescribed, the model based on Equation (1) together with a velocity field of the type (3)-(4) generates a Cauchy

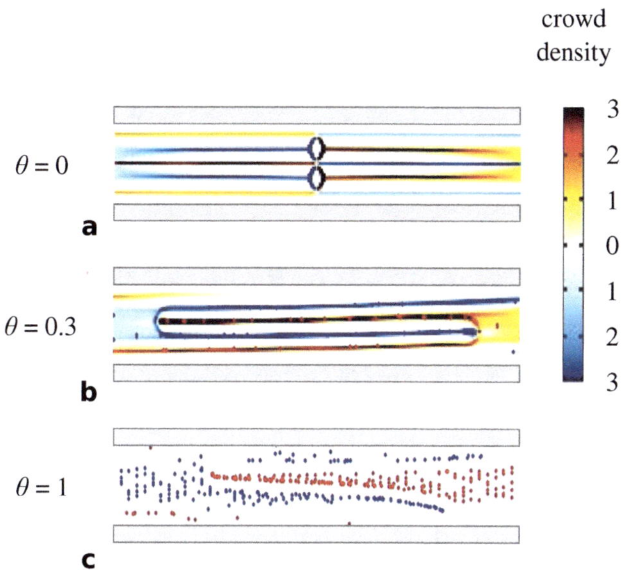

Figure 3. *Lane formation* in counter-flows obtained with model (1)-(3)-(4)-(9) with three different values of the perception parameter θ. **a.** With $\theta = 0$ pedestrian perception is purely *continuous*. Lanes emerge but the lack of *symmetry breaking* makes the result look rather artificial, thereby suggesting that granularity should play a role. **b.** With $\theta = 0.3$ pedestrian perception is genuinely *multiscale*. The atoms of μ_t introduce inhomogeneities in the density flow, which induce a qualitatively more realistic lane formation also at a purely continuous level (i.e., when looking at the density only). **c.** With $\theta = 1$ pedestrian perception is purely *discrete*. Also in this case lane-type patterns predicted by the model look realistic. On the whole, this example demonstrates that lane formation is quite a robust phenomenon at all scales, although pedestrian perception can influence the qualitative patterns collectively observed.

problem falling in the following class:

$$\begin{cases} \dfrac{\partial \mu_t}{\partial t} + \nabla \cdot (\mu_t v[\mu_t]) = 0 & x \in \mathbb{R}^d, \ t \in (0, T_{\max}] \\ \mu_0 = \bar{\mu}, \end{cases} \tag{11}$$

where we denoted by $v[\mu_t]$ a generic velocity field fully determined by the (unknown) measure μ_t.

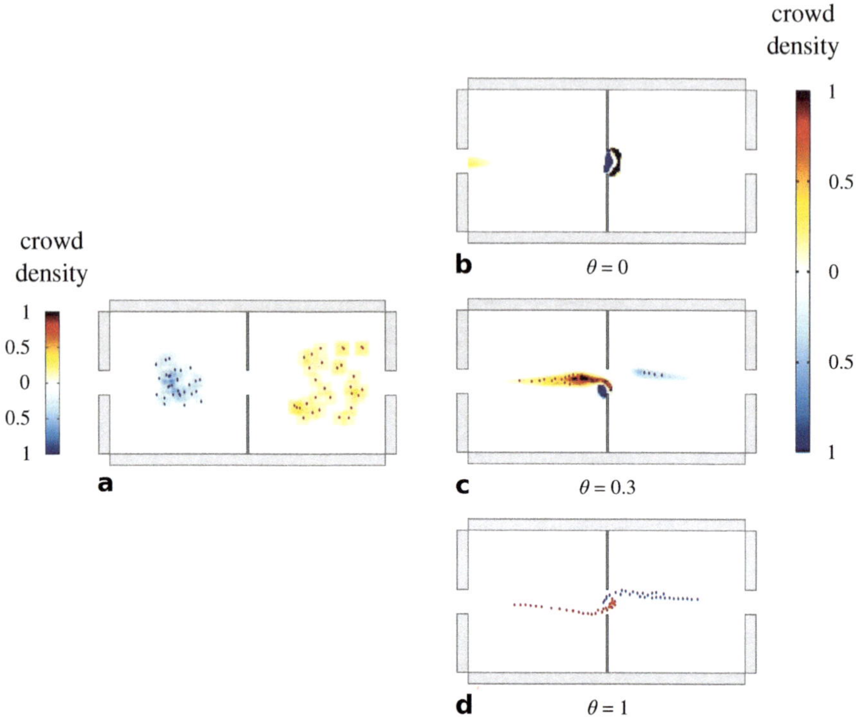

Figure 4. *Crossing flows at a bottleneck* obtained with model (1)-(3)-(4)-(9) with the same three values of the perception parameter θ as in Figure 3. **a.** Initial condition common to all cases. **b.** A purely *continuous* perception ($\theta = 0$) determines a *clogging* of the bottleneck, because individuals interact with subgroups of surrounding walkers being thus basically unable to exploit inter-pedestrian gaps. **c.** A genuinely *multiscale* perception (here with $\theta = 0.3$) produces instead a kind of *traffic light effect* at the bottleneck: the latter is occupied alternately by either crowd while the other one stops and waits. **d.** A purely *discrete* perception ($\theta = 1$) gives rise to an ordered *lane formation* through the bottleneck, for pedestrians estimate with great precision the position of nearby people and self-organize so as to share effectively the available room. On the whole, this example demonstrates that pedestrian perception can play a major role in shaping the observable collective patterns even starting from the very same initial conditions. Notice that all patterns shown in **b.**, **c.**, and **d.** can actually happen in real situations.

In general μ_t is a finite measure on $\mathbb{R}^d$ but not a probability measure. In fact we know that if $\bar{\mu}(\mathbb{R}^d) = N$ then $\mu_t(\mathbb{R}^d) = N$ for all $t \leq T_{\max}$ but clearly it has to be $N > 1$ in order for the model to describe interesting scenarios. However, for the analytical study of Problem (11) it is convenient to rescale μ_t with respect to the total number N of pedestrians in such a way that it is formally a probability, regardless of its derivation for modeling purposes. This way it is easier to set Problem (11) in the proper functional spaces with the proper metrics. Therefore we will henceforth assume to have implicitly performed such a rescaling (and we still denote by μ_t the rescaled measure)[1].

A proper weak sense in which Problem (11) can be understood is the one specified in (2) (with $\mu_0 = \bar{\mu}$). In particular, Equation (2) is well-defined if $t \mapsto \mu_t$ is a continuous mapping from the interval $(0, T_{\max}]$ into one of the metric spaces $\mathcal{P}_p(\mathbb{R}^d)$ of probability measures on $\mathbb{R}^d$ with finite p-th moment ($p \geq 1$, see Ambrosio et al. (2008) for technical details). Without loss of generality we fix $p = 1$, i.e., we consider the space $\mathcal{P}_1(\mathbb{R}^d)$, which is complete with the metric

$$W_1(\mu, \nu) = \sup_{\varphi \in \mathrm{Lip}_1(\mathbb{R}^d)} \int_{\mathbb{R}^d} \varphi \, d(\nu - \mu) \qquad (\mu, \nu \in \mathcal{P}_1(\mathbb{R}^d)),$$

called the *(first) Wasserstein distance.* In the definition above, $\mathrm{Lip}_1(\mathbb{R}^d)$ is the set of Lipschitz continuous functions on $\mathbb{R}^d$ whose Lipschitz constant is not greater than 1.

Ultimately, we say that:

Definition 4.1. A curve $\mu_\bullet \in C([0, T_{\max}]; \mathcal{P}_1(\mathbb{R}^d))$ is a (weak) solution to Problem (11) if it satisfies Equation (2), with $\mu_0 = \bar{\mu}$, for all $\phi \in C_c^\infty(\mathbb{R}^d)$ and all $t \in (0, T_{\max}]$.

The basic theory for Problem (11) depends essentially on the properties of the velocity field. We now formulate some assumptions, valid for sufficiently general fields v (i.e., not necessarily referred to the specific structure (3)-(4)), whence both the well-posedness of the Cauchy problem and the convergence of a suitable numerical scheme, based on the transport of measures, for the approximation of the solutions follow.

[1] Notice that if μ_t is thought of as a probability measure then the interaction velocity (4) must be coherently rewritten as

$$v_i[\mu_t](x) = N \int_{\mathbb{R}^d} K(x, y) \eta_{S(x)}(y) \, d\mu_t(y)$$

in view of the rescaling.

Assumptions on the velocity v for Problem (11)

(i) *Uniform boundedness*: there exists a constant $V_{\max} > 0$ such that

$$|v[\mu](x)| \leq V_{\max}, \qquad \forall\, x \in \mathbb{R}^d,\ \forall\, \mu \in \mathcal{P}_1(\mathbb{R}^d).$$

(ii) *Lipschitz continuity*: there exists a constant $\mathrm{Lip}(v) > 0$ such that

$$|v[\nu](y) - v[\mu](x)| \leq \mathrm{Lip}(v)\left(|y - x| + W_1(\mu,\, \nu)\right),$$

$\forall\, x,\, y \in \mathbb{R}^d,\ \forall\, \mu,\, \nu \in \mathcal{P}_1(\mathbb{R}^d)$.

(iii) *Mild linearity*: for all $\alpha \in [0,\, 1]$ and all pairs of measures $\mu,\, \nu \in \mathcal{P}_1(\mathbb{R}^d)$ it results

$$v[\alpha\mu + (1-\alpha)\nu] = \alpha v[\mu] + (1-\alpha)v[\nu].$$

Remark 4.2. We called Assumption (iii) *mild linearity* because it requires the mapping $\mu \mapsto v[\mu]$ to be linear for *convex* linear combinations only.

It is important to take into account that the Assumptions above are not meant to be sharp from the technical point of view. Rather, they are thought of for models which should be applied to realistic case studies. In this respect, one of their advantages is that they can be verified quite easily in practical cases. Furthermore, they allow for proofs which do not require sophisticated techniques of optimal transportation.

4.1 Well-posedness of Problem (11)

Using only Assumption (ii) it is possible to prove the following *a priori* estimate on the solutions to Problem (11):

Theorem 4.3. *If $\mu^1_\bullet,\, \mu^2_\bullet \in C([0,\, T_{\max}];\, \mathcal{P}_1(\mathbb{R}^d))$ are two solutions corresponding to two initial data $\bar{\mu}^1,\, \bar{\mu}^2 \in \mathcal{P}_1(\mathbb{R}^d)$, respectively, then there exists a constant $\mathcal{C} > 0$ such that*

$$W_1(\mu^1_t,\, \mu^2_t) \leq \mathcal{C} W_1(\bar{\mu}^1,\, \bar{\mu}^2), \qquad \forall\, t \in (0,\, T_{\max}]. \tag{12}$$

Proof. See e.g., Cristiani et al.. □

Thus the solution to Problem (11), if it exists, is unique and depends continuously on the initial datum. The constant $\mathcal{C}$ depends on the Lipschitz constant $\mathrm{Lip}(v)$ of the velocity and on the final time $T_{\max}$.

For the proof of the estimate (12) the interested reader is referred to the above-cited reference. Here we simply point out that the classical technique makes use of the following *representation formula* of the solutions to Problem (11): if $\mu_\bullet \in C([0, T_{\max}]; \mathcal{P}_1(\mathbb{R}^d))$ is a solution corresponding to an initial datum $\bar{\mu} \in \mathcal{P}_1(\mathbb{R}^d)$ then, after introducing the *flow map* $\gamma_t : \mathbb{R}^d \to \mathbb{R}^d$ defined by:

$$\begin{cases} \dfrac{\partial}{\partial t}\gamma_t(x) = v[\mu_t](\gamma_t(x)), & x \in \mathbb{R}^d,\ t \in (0, T_{\max}] \\ \gamma_0(x) = x, & x \in \mathbb{R}^d, \end{cases} \tag{13}$$

it results

$$\mu_t = \gamma_t \# \bar{\mu} \quad \text{viz.} \quad \mu_t(E) = \bar{\mu}(\gamma_t^{-1}(E)),\ \forall\, E \in \mathcal{B}(\mathbb{R}^d),$$

where $\#$ is the so-called *push forward* operator. This representation formula can be easily checked using Equation (2).

Next, in view of the further Assumptions (i), (iii), also existence of the solution can be proved:

Theorem 4.4. *For* $\bar{\mu} \in \mathcal{P}_2(\mathbb{R}^d) \subset \mathcal{P}_1(\mathbb{R}^d)$ *there exists a solution* $\mu_\bullet \in C([0, T_{\max}]; \mathcal{P}_1(\mathbb{R}^d))$ *to Problem* (11).

Proof. See e.g., Tosin and Frasca (2011). □

Notice that Theorem 4.4 requires actually $\bar{\mu} \in \mathcal{P}_2(\mathbb{R}^d)$, i.e., that the initial datum has both first and second order moments finite. This assumption is mainly technical. Nevertheless, from the point of view of applications it is not a limitation, since initial data typically have *compact* support (indeed, a crowd spread over the whole $\mathbb{R}^d$ would not be such a realistic initial condition), hence their moments of *any* order p are automatically finite. In fact:

$$\int_{\mathbb{R}^d} |x|^p \, d\bar{\mu}(x) = \int_{\operatorname{supp}(\bar{\mu})} |x|^p \, d\bar{\mu}(x) \leq R^p < +\infty,$$

R being the radius of one of the balls centered at the origin which contain $\operatorname{supp}(\bar{\mu})$.

Finally, we can establish the following well-posedness result, in the sense of Hadamard, for the Cauchy problem (11):

Theorem 4.5. *Let Assumptions (i)–(iii) hold. For all $\bar{\mu} \in \mathcal{P}_2(\mathbb{R}^d)$ there exists a unique solution $\mu_\bullet \in C([0, T_{\max}]; \mathcal{P}_1(\mathbb{R}^d))$ to Problem* (11) *in the sense of Definition 4.1. In addition, it depends continuously on the initial datum on the basis of the estimate* (12).

4.2 Numerical scheme and its convergence

In order to approximate the solution to Problem (11) it is possible to use the numerical scheme introduced by Piccoli and Tosin (2009, 2011), then further detailed for multiscale models by Cristiani et al. (2011) and adopted also by Maury et al. (2010); Canuto et al. (2012). In short, the idea is to approximate μ_t by an *absolutely continuous* measure with respect to Lebesgue, which is piecewise constant over a pairwise disjoint partition of $\mathbb{R}^d$ called *mesh.* Specifically, the latter is formed by cells E_i of characteristic size $h > 0$ such that $\mathcal{L}^d(E_i) \to 0$ when $h \to 0^+$:

$$\mathbb{R}^d = \bigcup_{i \in \mathbb{Z}^d} E_i, \qquad E_i \cap E_j = \emptyset \quad \forall i \neq j.$$

For $d = 2$ (which in several applications is the most interesting case for crowd simulations) the cells E_i can be square-shaped with edge length h. The simplest generalization to an arbitrary dimension is obtained by considering hypercubes with edge length h, so that $\mathcal{L}^d(E_i) = h^d$.

Let $\tilde{\mu}_n$ be the approximation of μ_t at the discrete time $t_n = n\Delta t$, where $\Delta t > 0$ is a fixed time step. Then by construction it results $d\tilde{\mu}_n = \tilde{\rho}_n\,dx$, where $\tilde{\rho}_n = \tilde{\rho}_n(x) : \mathbb{R}^d \to \mathbb{R}_+$ is the density that the numerical scheme has to determine. Again by construction, $\tilde{\rho}_n$ is piecewise constant on the grid $\{E_i\}_{i \in \mathbb{Z}^d}$, therefore it can be represented as

$$\tilde{\rho}_n(x) = \sum_{i \in \mathbb{Z}^d} \rho_i^n \chi_{E_i}(x), \tag{14}$$

χ_{E_i} denoting the characteristic function of the cell E_i. The unknowns are thus the coefficients $\rho_i^n \geq 0$.

The numerical scheme is constructed by imposing first that, in one time step, the measure $\tilde{\mu}_n$ is transported on the new measure $\tilde{\mu}_{n+1}$ by a suitable discretization $\tilde{\gamma}_n$ of the flow map (13): $\tilde{\mu}_{n+1} = \tilde{\gamma}_n \# \tilde{\mu}_n$, and by testing then this relation on the grid cells: $\tilde{\mu}_{n+1}(E_i) = \tilde{\mu}_n(\tilde{\gamma}_n^{-1}(E_i))$. Using the numerical density (14), this yields a recursive formula to pass from the coefficients ρ_i^n to those at the next time step:

$$\rho_i^{n+1} = \frac{1}{h^d} \sum_{j \in \mathbb{Z}^d} \rho_j^n \mathcal{L}^d(E_i \cap \tilde{\gamma}_n(E_j)) \quad (i \in \mathbb{Z}^d). \tag{15}$$

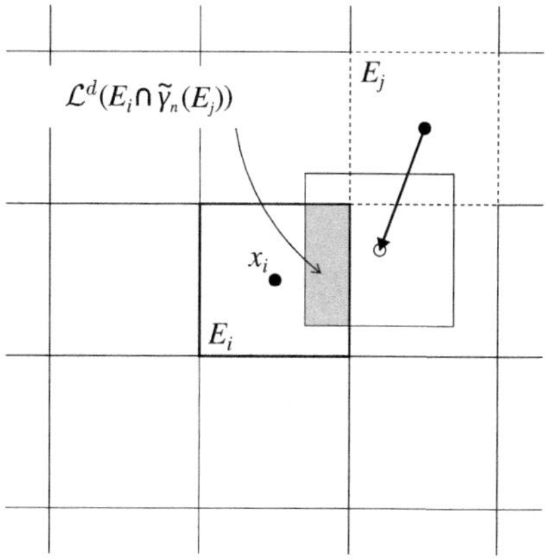

Figure 5. Action of the numerical scheme (15) on the grid cells.

It expresses the fact that the numerical density is redistributed over the mesh, in one time step, proportionally to the (Lebesgue) measure of the intersections among the cells moved by the discrete flow map $\tilde{\gamma}_n$ (see Figure 5). In particular, the latter is obtained cell-by-cell as:

$$\tilde{\gamma}_n(x) = x + v[\tilde{\mu}_n](x_i)\Delta t \quad \text{for } x \in E_i,$$

where x_i is any point of E_i, for instance its center. Hence $\tilde{\gamma}_n$ acts in every cell as a rigid translation with constant velocity $v[\tilde{\mu}_n](x_i)$, namely the velocity v of the exact problem (11) computed for $x = x_i$ with respect to the approximate measure $\tilde{\mu}_n$.

Despite that the numerical measure $\tilde{\mu}_n$ has been chosen absolutely continuous with respect to Lebesgue, the scheme (15) can actually approximate solutions to Problem (11) with generic spatial structure. Indeed the following result holds true:

Theorem 4.6. *Consider a sequence of spatiotemporal grids indexed by* $k = 0, 1, 2 \ldots$*, with mesh parameters* h_k, $\Delta t_k \to 0$ *for* $k \to \infty$*. Let* $\tilde{\mu}_t^k$ *be the piecewise linear interpolation in time of the numerical solutions* $\tilde{\mu}_n^k$ *computed on the k-th mesh with the scheme* (15)*:*

$$\tilde{\mu}_t^k = \sum_{n=0}^{N_{\max}^k - 1} \left[\left(1 - \frac{t - t_n^k}{\Delta t_k}\right) \tilde{\mu}_n^k + \frac{t - t_n^k}{\Delta t_k} \tilde{\mu}_{n+1}^k \right] \chi_{[t_n^k, t_{n+1}^k]}(t), \qquad (16)$$

where $t_n^k = n\Delta t_k$, $N_{\max}^k$ *is the number of discrete time steps on the k-th grid needed to reach the final time* $T_{\max}$, *and* $\chi_{[t_n^k,\, t_{n+1}^k]}$ *denotes the characteristic function of the interval* $[t_n^k,\, t_{n+1}^k]$.

Assume that Assumptions (i)–(iii) on page 12 hold true, that the initial datum is discretized as:

$$(\rho_i^0)^k = \frac{\bar{\mu}(E_i^k)}{h_k^d} \qquad (i \in \mathbb{Z}^d),$$

and moreover that the spatiotemporal grids are chosen in such a way that

$$h_k = o(\Delta t_k) \quad \text{when} \quad k \to \infty.$$

If $\tilde{\mu}_\bullet^k$ *converges in* $C([0,\, T_{\max}];\, \mathcal{P}_1(\mathbb{R}^d))$ *to some* $\mu_\bullet$ *for* $k \to \infty$ *in the following sense:*

$$\lim_{k\to\infty} \sup_{t\in[0,\, T_{\max}]} W_1(\tilde{\mu}_t^k,\, \mu_t) = 0$$

then the limit $\mu_\bullet$ *is a (weak) solution to Problem* (11) *in the sense of Definition 4.1.*

Proof. See e.g., Tosin and Frasca (2011). □

Notice that Theorem 4.6 does not guarantee but *assumes* the convergence of the numerical solution to a curve $\mu_\bullet \in C([0,\, T_{\max}];\, \mathcal{P}_1(\mathbb{R}^d))$. For this reason, it recalls the Lax-Wendroff's Theorem about the convergence of numerical schemes for hyperbolic conservation laws. Nevertheless it is possible to complement it with a simple criterion ensuring that the required convergence does indeed take place:

Proposition 4.7. *Let a compact set* $K \subset \mathbb{R}^d$ *exist such that* $\operatorname{supp}(\tilde{\mu}_n^k) \subseteq K$ *for all n and all k. Then the time-interpolated sequence* $\{\tilde{\mu}_\bullet^k\}_{k\geq 0}$, *cf. Equation* (16), *converges in* $C([0,\, T_{\max}];\, \mathcal{P}_1(\mathbb{R}^d))$.

Proof. See e.g., Tosin and Frasca (2011). □

It is evident that such a K exists especially when the initial datum $\bar{\mu}$ is compactly supported. In fact, due to Assumption (i), Problem (11) describes a transport with *finite* speed, hence $\operatorname{supp}(\bar{\mu})$ cannot expand indefinitely within a finite time $T_{\max}$.

Besides the references already given for the proofs of Theorem 4.6 and of Proposition 4.7, the interested reader is referred also to the paper by Piccoli and Rossi (2013) for a deep analysis of further numerical schemes dealing with the approximation of the solution to Problem (11).

4.3 Structure of the solution

The theory set forth in the previous sections does not provide any information about the spatial structure of the solutions to Problem (11), which is important especially for multiscale models discussed in Section 3.

Using the representation formula introduced in Section 4.1, it is quite easy to see that if the initial datum $\bar{\mu}$ has a discrete structure like[2] $\bar{\mu} = \frac{1}{N}\sum_{i=1}^{N} \delta_{\bar{x}_i}$ then also the solution μ_t has a discrete structure for all $t > 0$. In fact, for all Borel set $E \in \mathcal{B}(\mathbb{R}^d)$ it results:

$$\mu_t(E) = (\gamma_t\#\bar{\mu})(E) = \bar{\mu}(\gamma_t^{-1}(E)) = \frac{1}{N}\sum_{i=1}^{N} \delta_{\bar{x}_i}(\gamma_t^{-1}(E)) = \frac{1}{N}\sum_{i=1}^{N} \delta_{\gamma_t(\bar{x}_i)}(E),$$

hence letting $x_i(t) := \gamma_t(\bar{x}_i)$ we have $\mu_t = \frac{1}{N}\sum_{i=1}^{N} \delta_{x_i(t)}$. Notice that this holds independently of the regularity of the flow map γ_t, namely of the velocity field $v[\mu_t]$.

Conversely, for an absolutely continuous initial datum, $d\bar{\mu}(x) = \bar{\rho}(x)\,dx$, the solution μ_t may develop singularities in finite time if the flow map tends to concentrate "too much" density over "too small" spatial structures. The following result gives a sufficient condition for ruling out this possibility:

Theorem 4.8. *Let $\bar{\mu}$ be absolutely continuous with respect to the Lebesgue measure. Assume that at every fixed time $t \in (0, T_{\max}]$ there exists a constant $C_t > 0$, possibly depending on t, such that*

$$\mathcal{L}^d(\gamma_t^{-1}(E)) \le C_t \mathcal{L}^d(E), \qquad \forall\, E \in \mathcal{B}(\mathbb{R}^d). \tag{17}$$

Then also $\mu_t = \gamma_t\#\bar{\mu}$ is absolutely continuous with respect to Lebesgue for all $t \in (0, T_{\max}]$.

Proof. See e.g., Cristiani et al. (2011); Piccoli and Tosin (2011); Cristiani et al.. □

Condition (17) requires that the flow map does not shrink measurable sets too much. Indeed, by reading the inequality from right to left, it states that the Lebesgue measure of E can be controlled *from below* by the measure of its inverse image through γ_t. Nevertheless, this condition is not easy to be checked in concrete cases. To obviate such a difficulty, at least for smooth flow maps, it is possible to use the following criterion, which is at the same time sufficient and easier to verify.

[2]Unlike Equation (5), here the coefficient $\frac{1}{N}$ appears because of the rescaling to a probability measure introduced at the beginning of the section.

Proposition 4.9. *If the flow map* γ_t *is a diffeomorphism with Lipschitz constant* $\mathrm{Lip}(\gamma_t)$ *such that*

$$\mathrm{Lip}(\gamma_t) < \frac{1}{\mathrm{Lip}(v)T_{\max}} \tag{18}$$

then it fulfills condition (17).

Proof. See e.g., Cristiani et al.. □

In particular, it is useful to know that under Assumption (ii) the Lipschitz constant of γ_t can be estimated as $\mathrm{Lip}(\gamma_t) \leq 1 + \mathrm{Lip}(v)T_{\max}e^{\mathrm{Lip}(v)T_{\max}}$ (see Cristiani et al.), hence (18) is certainly satisfied if

$$1 + \mathrm{Lip}(v)T_{\max}e^{\mathrm{Lip}(v)T_{\max}} < \frac{1}{\mathrm{Lip}(v)T_{\max}},$$

which is ultimately a condition on the Lipschitz constant of the velocity. This is more practical because, as seen in Section 2.1, it is the velocity, not the flow map, which plays a major role in the modeling approach.

Finally, if the initial datum has a hybrid structure such as (9) it is sufficient to recall the linearity of the push forward operator # to conclude that the results above apply separately to the discrete and continuous parts. Consequently, if the flow map satisfies the conditions expressed by Theorem 4.8 and by Proposition 4.9 then the solution to Problem (11) preserves the multiscale structure for all times $t > 0$.

5 Back to crowd models

Results presented in Section 4 hold for an "abstract" velocity field characterized essentially by Assumptions (i)–(iii) (cf. page 12). In order to construct crowd models not only physically realistic but also mathematically robust it is therefore important to study the interplay between the structures introduced in Section 2 and these assumptions.

As recalled in Section 2.1, the interaction kernel K should depend on the *relative position* $y - x$ of the interacting pedestrians in order for the description of the interactions to be independent of rigid transformations of the coordinate system. This implies

$$K(x, y) = k(y - x),$$

$k : \mathbb{R}^d \to \mathbb{R}^d$ being a function to be properly modeled (which we still call *interaction kernel*). That said, pedestrian velocity (3)-(4) takes the form:

$$v[\mu_t](x) = v_{\mathrm{d}}(x) + N\int_{\mathbb{R}^d} k(y - x)\eta_{S(x)}(y)\,d\mu_t(t), \tag{19}$$

where the coefficient N (total number of pedestrians) in front of the interaction integral appears because of the reinterpretation of μ_t as a probability measure (cf. footnote 1 on page 11).

It is immediate to check that the velocity field (19) satisfies Assumption (iii) with no additional hypotheses. In fact, it is sufficient to write $v_{\mathrm{d}} = \alpha v_{\mathrm{d}} + (1 - \alpha) v_{\mathrm{d}}$ and to collect the terms conveniently. Notice that it is fundamental that Assumption (iii) requires only a *mild* linearity, for the mapping $\mu \mapsto v[\mu]$ resulting from Equation (19) is in general not linear (except when the desired velocity is zero, which however does not always make sense from the modeling point of view).

In order for the velocity (19) to fulfill also Assumptions (i), (ii) some further technical details are needed, to be regarded as modeling guidelines, concerning the structure of the terms v_{d}, k, $\mathcal{S}(\cdot)$, and $\eta_{\mathcal{S}(\cdot)}$. For the sake of simplicity we consider only the two-dimensional case ($d = 2$), which is however largely sufficient for addressing realistic crowd models.

Modeling Guidelines for the velocity v (19) with $d = 2$

(i) *Desired velocity*: let $x \mapsto v_{\mathrm{d}}(x)$ be Lipschitz continuous and bounded in $\mathbb{R}^2$.

(ii) *Sensory region*: for all $x \in \mathbb{R}^2$, let $\mathcal{S}(x)$ be a bounded Borel set contained in a ball with fixed radius $R > 0$ independent of x (for instance, the one centered in x: $\mathcal{S}(x) \subseteq B_R(x)$) and isometric to a reference set $\mathcal{S}(0) \subseteq B_R(0)$.

(iii) *Interaction kernel*: let $x \mapsto k(x)$ be Lipschitz continuous in the ball $B_R(0)$ centered at the origin and with radius R defined at the previous point (ii).

(iv) *Cut-off function*: for all $E \in \mathcal{B}(\mathbb{R}^2)$, let $x \mapsto \eta_E(x)$ be Lipschitz continuous and bounded between 0 and 1 in $\mathbb{R}^2$ with $\mathrm{supp}(\eta_E) \subset\subset E$.

Remark 5.1. The Modeling Guideline (ii) means that the sensory region $\mathcal{S}(x)$ of the point x is obtained by translating and rotating the reference set $\mathcal{S}(0)$. The translation vector is obviously x. Moreover, according to what has been said in Section 2.1, the rotation angle is individuated by the direction of the vector $v_{\mathrm{d}}(x)$.

If pedestrian velocity is constructed in accordance with the Modeling Guidelines stated above then the crowd model based on Equations (1), (19) features a certain mathematical robustness, indeed:

Proposition 5.2. *If, in two space dimensions ($d = 2$), the velocity field* (19) *complies with the Modeling Guidelines (i)–(iv) then it fulfills Assumptions (i)–(iii).*

Proof. See e.g., Tosin and Frasca (2011). □

hence it is possible to apply to it the theory of well-posedness and numerical approximation presented in Section 4.

Bibliography

L. Ambrosio, N. Gigli, and G. Savaré. *Gradient flows in metric spaces and in the space of probability measures.* Lectures in Mathematics ETH Zürich. Birkhäuser Verlag, Basel, second edition, 2008.

L. Bruno, A. Tosin, P. Tricerri, and F. Venuti. Non-local first-order modelling of crowd dynamics: A multidimensional framework with applications. *Appl. Math. Model.*, 35(1):426–445, 2011.

C. Canuto, F. Fagnani, and P. Tilli. An Eulerian approach to the analysis of Krause's consensus models. *SIAM J. Control Optim.*, 50(1):243–265, 2012.

R. M. Colombo and M. D. Rosini. Existence of nonclassical solutions in a pedestrian flow model. *Nonlinear Anal. Real World Appl.*, 10(5):2716–2728, 2009.

V. Coscia and C. Canavesio. First-order macroscopic modelling of human crowd dynamics. *Math. Models Methods Appl. Sci.*, 18(1 suppl.):1217–1247, 2008.

E. Cristiani, B. Piccoli, and A. Tosin. Multiscale Modeling of Pedestrian Dynamics. In preparation.

E. Cristiani, B. Piccoli, and A. Tosin. Modeling self-organization in pedestrians and animal groups from macroscopic and microscopic viewpoints. In G. Naldi, L. Pareschi, and G. Toscani, editors, *Mathematical Modeling of Collective Behavior in Socio-Economic and Life Sciences*, Modeling and Simulation in Science, Engineering and Technology, pages 337–364. Birkhäuser, Boston, 2010.

E. Cristiani, B. Piccoli, and A. Tosin. Multiscale modeling of granular flows with application to crowd dynamics. *Multiscale Model. Simul.*, 9 (1):155–182, 2011.

E. Cristiani, B. Piccoli, and A. Tosin. How can macroscopic models reveal self-organization in traffic flow? In *Proceedings of the 51st IEEE Conference on Decision and Control*, Maui, Hawaii, USA, December 2012.

B. Maury, A. Roudneff-Chupin, and F. Santambrogio. A macroscopic crowd motion model of gradient flow type. *Math. Models Methods Appl. Sci.*, 20(10):1787–1821, 2010.

B. Piccoli and F. Rossi. Transport equation with nonlocal velocity in Wasserstein spaces: convergence of numerical schemes. *Acta Appl. Math.*, 124(1):73–105, 2013.

B. Piccoli and A. Tosin. Pedestrian flows in bounded domains with obstacles. *Contin. Mech. Thermodyn.*, 21(2):85–107, 2009.

B. Piccoli and A. Tosin. Time-evolving measures and macroscopic modeling of pedestrian flow. *Arch. Ration. Mech. Anal.*, 199(3):707–738, 2011.

E. Schröedinger. *What is Life? Mind and Matter.* Cambridge University Press, 1967.

A. Tosin and P. Frasca. Existence and approximation of probability measure solutions to models of collective behaviors. *Netw. Heterog. Media*, 6(3): 561–596, 2011.

GPSR Compliance

The European Union's (EU) General Product Safety Regulation (GPSR) is a set of rules that requires consumer products to be safe and our obligations to ensure this.

If you have any concerns about our products, you can contact us on ProductSafety@springernature.com

In case Publisher is established outside the EU, the EU authorized representative is:

Springer Nature Customer Service Center GmbH
Europaplatz 3
69115 Heidelberg, Germany

Batch number: 10732220

Printed by Printforce, the Netherlands